Current Advances on Non-Melanoma Skin Cancer

Current Advances on Non-Melanoma Skin Cancer

Editor

Constantin Caruntu

MDPI • Basel • Beijing • Wuhan • Barcelona • Belgrade • Manchester • Tokyo • Cluj • Tianjin

Editor
Constantin Caruntu
"Carol Davila" University of
Medicine and Pharmacy
Romania

Editorial Office
MDPI
St. Alban-Anlage 66
4052 Basel, Switzerland

This is a reprint of articles from the Special Issue published online in the open access journal *Journal of Clinical Medicine* (ISSN 2077-0383) (available at: https://www.mdpi.com/journal/jcm/special_issues/Non-Melanoma_Skin_Cancer).

For citation purposes, cite each article independently as indicated on the article page online and as indicated below:

LastName, A.A.; LastName, B.B.; LastName, C.C. Article Title. *Journal Name* **Year**, *Volume Number*, Page Range.

ISBN 978-3-0365-0890-0 (Hbk)
ISBN 978-3-0365-0891-7 (PDF)

© 2021 by the authors. Articles in this book are Open Access and distributed under the Creative Commons Attribution (CC BY) license, which allows users to download, copy and build upon published articles, as long as the author and publisher are properly credited, which ensures maximum dissemination and a wider impact of our publications.

The book as a whole is distributed by MDPI under the terms and conditions of the Creative Commons license CC BY-NC-ND.

Contents

About the Editor vii

Preface to "Current Advances on Non-Melanoma Skin Cancer" ix

Taxiarchis Konstantinos Nikolouzakis, Luca Falzone, Konstantinos Lasithiotakis, Sabine Krüger-Krasagakis, Alexandra Kalogeraki, Maria Sifaki, Demetrios A. Spandidos, Emmanuel Chrysos, Aristidis Tsatsakis and John Tsiaoussis
Current and Future Trends in Molecular Biomarkers for Diagnostic, Prognostic, and Predictive Purposes in Non-Melanoma Skin Cancer
Reprinted from: *J. Clin. Med.* **2020**, *9*, 2868, doi:10.3390/jcm9092868 1

Mircea Tampa, Simona Roxana Georgescu, Cristina Iulia Mitran, Madalina Irina Mitran, Clara Matei, Cristian Scheau, Carolina Constantin and Monica Neagu
Recent Advances in Signaling Pathways Comprehension as Carcinogenesis Triggers in Basal Cell Carcinoma
Reprinted from: *J. Clin. Med.* **2020**, *9*, 3010, doi:10.3390/jcm9093010 23

Paola Pasquali, Gonzalo Segurado-Miravalles, Mar Castillo, Ángeles Fortuño, Susana Puig and Salvador González
Use of Cytology in the Diagnosis of Basal Cell Carcinoma Subtypes
Reprinted from: *J. Clin. Med.* **2020**, *9*, 612, doi:10.3390/jcm9030612 39

Mihai Lupu, Iris Maria Popa, Vlad Mihai Voiculescu, Ana Caruntu and Constantin Caruntu
A Systematic Review and Meta-Analysis of the Accuracy of in Vivo Reflectance Confocal Microscopy for the Diagnosis of Primary Basal Cell Carcinoma
Reprinted from: *J. Clin. Med.* **2019**, *8*, 1462, doi:10.3390/jcm8091462 49

Andreea D. Lazar, Sorina Dinescu and Marieta Costache
Deciphering the Molecular Landscape of Cutaneous Squamous Cell Carcinoma for Better Diagnosis and Treatment
Reprinted from: *J. Clin. Med.* **2020**, *9*, 2228, doi:10.3390/jcm9072228 67

Ricardo Moreno, Laura Nájera, Marta Mascaraque, Ángeles Juarranz, Salvador González and Yolanda Gilaberte
Influence of Serum Vitamin D Level in the Response of Actinic Keratosis to Photodynamic Therapy with Methylaminolevulinate
Reprinted from: *J. Clin. Med.* **2020**, *9*, 398, doi:10.3390/jcm9020398 91

Mihai Lupu, Ana Caruntu, Daniel Boda and Constantin Caruntu
In Vivo Reflectance Confocal Microscopy-Diagnostic Criteria for Actinic Cheilitis and Squamous Cell Carcinoma of the Lip
Reprinted from: *J. Clin. Med.* **2020**, *9*, 1987, doi:10.3390/jcm9061987 101

About the Editor

Constantin Caruntu (MD, Ph.D.) is professor of Physiology at the "Carol Davila" University of Medicine and Pharmacy and is also a dermatologist at the "Prof. N. Paulescu" National Institute of Diabetes in Bucharest. He is trained in in vivo reflectance confocal microscopy, cellular and molecular biology techniques, and molecular imaging techniques. He is member of the International Dermoscopy Society, International League of Dermatological Societies, International Union of Physiological Sciences and many other national or international scientific societies. Furthermore, he is academic editor, guest editor and reviewer for various prestigious scientific journals. His research interests include in vivo reflectance confocal microscopy, dermato-oncology, inflammatory skin diseases, neurogenic inflammation and neuroendocrinology of the skin.

Preface to "Current Advances on Non-Melanoma Skin Cancer"

Non-melanoma skin cancer (NMSC) comprises basal cell carcinoma (BCC), squamous cell carcinoma (SCC), and several rare skin tumors and is the most common malignancy that affects humans worldwide. It accounts for the vast majority of skin cancers and a large percentage of all malignant tumors. Despite the growing public awareness and scientific interest regarding the risk of skin cancer, the incidence of NMSC is still rapidly increasing. Even if most NMSCs are associated with a less aggressive behavior, they can still be locally invasive and may produce extensive destruction of neighboring structures, inducing significant morbidity. Moreover, different types or subtypes of NMSCs are associated with frequent recurrence and may carry a significant metastatic potential. As a direct consequence, NMSC has become a major burden on healthcare systems with a significant socio-economic impact. Hence, there is no doubt as to why these keratinocyte-derived tumors are in the spotlight of scientific interest for both fundamental research and clinical practice. Consequently, this Special Issue brings together recent and relevant scientific research on NMSC. Studies regarding the complex inherited and environmental factors that can trigger tumor initiation and progression may offer a new perspective on skin cancer prevention. Investigation of new markers of skin carcinogenesis may contribute to the development of novel diagnostic, staging, and prognosis strategies for skin cancer and could lead to the design of more sophisticated and individually tailored treatment protocols. Improvements related to non-invasive or minimally invasive diagnostic tools and treatment methods may ameliorate the discomfort of patients while also reducing the costs associated with therapy. Moreover, the development of new techniques for the early diagnosis of NMSCs could reduce morbidity, ensuring more efficient treatment with better aesthetic and functional results.

Constantin Caruntu
Editor

Review

Current and Future Trends in Molecular Biomarkers for Diagnostic, Prognostic, and Predictive Purposes in Non-Melanoma Skin Cancer

Taxiarchis Konstantinos Nikolouzakis [1,2], Luca Falzone [3], Konstantinos Lasithiotakis [2], Sabine Krüger-Krasagakis [4], Alexandra Kalogeraki [5], Maria Sifaki [6], Demetrios A. Spandidos [7], Emmanuel Chrysos [2], Aristidis Tsatsakis [6,*] and John Tsiaoussis [1,*]

1. Laboratory of Anatomy-Histology-Embryology, Medical School, University of Crete, 71110 Heraklion, Crete, Greece; medp2011826@med.uoc.gr
2. Department of General Surgery, University General Hospital of Heraklion, 71110 Heraklion, Crete, Greece; k.lasithiotakis@uoc.gr (K.L.); manolischrysos@gmail.com (E.C.)
3. Epidemiology Unit, IRCCS Istituto Nazionale Tumori 'Fondazione G. Pascale', I-80131 Naples, Italy; l.falzone@istitutotumori.na.it
4. Dermatology Department, University Hospital of Heraklion, 71110 Heraklion, Crete, Greece; krkras@med.uoc.gr
5. Department of Pathology-Cytopathology, Medical School, University of Crete, 70013 Heraklion, Crete, Greece; a.kalogeraki@med.uoc.gr
6. Centre of Toxicology Science and Research, Faculty of Medicine, University of Crete, 71003 Heraklion, Crete, Greece; skinclinicsifaki@gmail.com
7. Laboratory of Clinical Virology, Medical School, University of Crete, 71003 Heraklion, Crete, Greece; spandidos@spandidos.gr
* Correspondence: tsatsaka@uoc.gr (A.T.); tsiaoussis@uoc.gr (J.T.)

Received: 3 August 2020; Accepted: 1 September 2020; Published: 4 September 2020

Abstract: Skin cancer represents the most common type of cancer among Caucasians and presents in two main forms: melanoma and non-melanoma skin cancer (NMSC). NMSC is an umbrella term, under which basal cell carcinoma (BCC), squamous cell carcinoma (SCC), and Merkel cell carcinoma (MCC) are found along with the pre-neoplastic lesions, Bowen disease (BD) and actinic keratosis (AK). Due to the mild nature of the majority of NMSC cases, research regarding their biology has attracted much less attention. Nonetheless, NMSC can bear unfavorable characteristics for the patient, such as invasiveness, local recurrence and distant metastases. In addition, late diagnosis is relatively common for a number of cases of NMSC due to the inability to recognize such cases. Recognizing the need for clinically and economically efficient modes of diagnosis, staging, and prognosis, the present review discusses the main etiological and pathological features of NMSC as well as the new and promising molecular biomarkers available including telomere length (TL), telomerase activity (TA), CpG island methylation (CIM), histone methylation and acetylation, microRNAs (miRNAs), and micronuclei frequency (MNf). The evaluation of all these aspects is important for the correct management of NMSC; therefore, the current review aims to assist future studies interested in exploring the diagnostic and prognostic potential of molecular biomarkers for these entities.

Keywords: non-melanoma skin cancer; basal cell carcinoma; squamous cell carcinoma; Merkel cell carcinoma; telomeres; telomerase; epigenetics; miRNA

1. Introduction

Skin cancer is currently the most common type of cancer among Caucasians [1]. It is estimated that approximately 1 in 5 Americans will develop skin cancer at some point in their lives by the age of

70 [2]. Unfortunately, in spite of immense efforts being made in public health awareness and primary prevention campaigns, a steady increase in skin cancer rates is observed [3–5]. In fact, non-melanoma skin cancer (NMSC) is the most common type with a relative incidence increase of up to 10% per annum, with 2–3 million new cases each year globally [6]. Skin cancer includes several distinct subtypes which can be divided in two main categories, malignant melanoma, and NMSC, with the latter being further divided into basal cell carcinoma (BCC), cutaneous squamous cell carcinoma (cSCC), Bowen disease (BD), actinic keratosis (AK), and Merkel cell carcinoma (MCC) each of which has a different biological behavior, etiology, and prognosis [6]. From these five distinct entities, BCC, SCC, and MCC stand out, given their potential to invade into deeper layers and metastasize [7,8].

BD is in nature an in situ SCC, while AK is a precancerous lesion acting as precursor to SCC. Even though they both exhibit a close association with SCC, they present different histopathological findings [8]. BCCs are more benign lesions having an almost absent metastatic potential, whereas SCCs exhibit a metastatic risk between 0.1–13.7% [9]. Given the fact that the global population is aging, an increase in the associated morbidity and local recurrence rates is to be expected. This in hand creates a great burden on national healthcare systems and economies. Accounting for 70–80% of all skin cancer cases, BCC is ranked among the most common types of cancer [10]. However, given the benign nature of BCCs and the ease of treatment in a doctors' office, the majority of cases are not recorded in most national cancer registries [11]. BCC preferentially arises from stem cells within hair follicles and mechanosensory niches [12]. Generally, BCC is a slow-growing tumor which rarely gives rise to distant metastases. However, if left untreated, it can grow invasively, destroying underlying tissues. It has been shown that patients with BCC face a 10-fold risk of developing another BCC compared to the general population [13]. Nonetheless, given its benign character, no long-term follow-up is required following a complete resection of the primary tumor [14]. SCC is the second most frequent type of skin cancer [3]. As already mentioned, SCC usually occurs on sun-exposed areas of the skin, such as the head, face, earlobe, lips, or torso. Nonetheless, it can also arise from the surrounding skin of the anus and genitalia, or even from skin with chronic inflammation, such as a scar or chronic wound [15]. If left untreated, an in situ SCC (AK or BD) may evolve into an invasive SCC with a great risk of metastasizing or relapsing [16].

MCC is a rare type of NMSC arising from Merkel cells. Epidemiological findings have identified UV radiation, old age, male sex, and Caucasian descent as strong risk factors contributing to the surprising increase in incidence rates by 95% between 2000 and 2013 [17]. In cases with immunosuppression in particular, an aggressive form is exhibited with mortality rates exceeding 30% [18–21]. However, the pathophysiology of MCC development is not yet fully understood. Under poorly understood circumstances, Merkel cells produce the neuroendocrine lesion termed MCC. From its early discovery in 1972 by Toker [22], MCC has changed several names some of which are "cutaneous neuroendocrine carcinoma", "cutaneous trabecular carcinoma", and "small cell primary cutaneous carcinoma" [23]. Various mechanisms have been suggested to induce Merkel cell carcinogenesis. such as cellular senescence, immunosuppression, and the potential oncogenic pathways induced by UV exposure (UV-specific mutations in the p53 gene) [24]. Recently, Feng et al. found that a novel type of polyoma virus may attribute to MCC formation [25], highlighting the increased complexity of this entity. Being a multifactorial disease, NMSC remains a challenge for clinicians and researchers, not only to understand its biological behavior, but also to develop better tailored and personalized treatment plans [26]. Fortunately, as proven from various national cancer registries, the majority of NMSC cases exhibit an excellent 5-year survival rate ranging from 100% for BCC to 95% for SCC [27,28] with local recurrence rates being <5% [29,30]. It is clear that the early diagnosis of primary and relapsed tumors in addition to carefully tailored treatments will be greatly assisted from the introduction of appropriate biomarker panels into everyday clinical practice. Thus, the present brief review aims not only to introduce the clinical significance of using biomarkers for NMSC, but also to pinpoint novel biomarkers worthy of further research. From the great number of molecular biomarkers under research, we choose to present those which are most likely to be introduced into everyday clinical practice in the near future, such as the miRNAs, and those which according to the current literature are most promising candidates requiring further investigation.

2. Etiology

According to the current knowledge on NMSC development, a constellation of factors are found to be implicated such as environmental exposure to UV light (regions closer to the equator suffer from higher rates of NMSC) [5,31], radiotherapy [32], viral infections (mostly β-HPV) [3], immunosuppression (based primarily upon the increased incidence exhibited in organ transplant recipients and the twofold higher incidence rate among HIV+ patients where SCCs is positively correlated with immunosuppression) [33,34], and genetic predisposition [35].

2.1. Ultraviolet (UV) Light

UV light exposure has been found to result in DNA mutations by inducing covalent bonding between adjacent pyrimidines (from UVB) and the formation of reactive oxygen species (from UVA) [36]. In detail, NMSCs formation has been positively associated to recreational UV light exposure with 2.5- and 1.5-fold increase in the risk of developing SCC and BCC, respectively [37]. Moreover, prolonged sunlight exposure during childhood and adolescence has been found to be responsible for BCCs, while chronic UV exposure is SCC formation in more advanced ages [1]. Notably, UV light may have a carcinogenic effect via immunosuppression. In detail, it has been described that a cellular modulation of immune cells is evoked, as evidenced by the concomitant depletion of Langerhans cells from the epidermis, altered antigen presentation in the lymph nodes, a shift towards Th2 responses and the development of tumor antigen-specific T regulatory cells, resulting in blocked immune surveillance and tumor outgrowth [38–40].

2.2. Genetic Background

Genetic predisposition is neither present nor uniform across all NMSCs. Most BCCs lack any pre-existing genetic background while SCCs may arise from a genetically predisposed clonal cell growth. Genetic damage accumulates, leading first to precursor lesions of AK or BD and subsequently to SCC [6] allowing even for multifocal development of SCCs (field cancerization) [41,42]. Several tumor suppressor genes and proto-oncogenes have been found to be implicated in BCC pathogenesis, such as components of the Sonic Hedgehog pathway (PTCH1 and SMO), the TP53 tumor suppressor gene, and members of the RAS family. In fact, it seems that the improper activation of the Sonic Hedgehog pathway is the key component pathway in BCC carcinogenesis [43,44]. SCCs are also driven by several mutated genes [45]. In detail, several mutations of the tyrosine kinase receptors (epidermal growth factor receptor-EGFR and fibroblast growth factor receptors—FGFRs) [46], certain cell cycle regulatory genes (TP53-the most common somatic mutation, CDKN2A/RB1, CCDN1, and MYC) [47,48], the RAS/MAPK and PI3K signaling pathways [46], genomic loci implicated in squamous cell fate determination (TP63, SOX2, and NRF2) [49–51], and squamous differentiation network (Notch and Fat1) [52,53] have been found.

2.3. Infectious Agents

An increasing body of evidence highlights the oncogenic potential of certain viruses such as the HPV, EBV, and the recently discovered Merkel Cell Polyomavirus (MCPyV) for NMSCs. HPV produces the E6 and E7 oncoproteins which have the potential to integrate into the hosts' keratinocytes genome [54,55]. It is worth noting that HPV-positive NMSC presents a more benign clinical behavior than HPV-negative NMSC. Even though the reason behind this remains undetermined, it may be due to the fact that the majority of the HPV-positive NMSCs tend to express wild-type TP53. On the counterpart, the majority of the HPV-negative cases exhibit mutated TP53 with or without accompanying mutations in other genomic loci [45]. On the contrary, EBV-induced carcinogenesis results from a multistep process, where the effect from a chronic EBV infection augments the results driven from genetic and epigenetic (methylation of several genomic sites and modulators) changes in the keratinocytes' genome [56]. In 2008, Feng et al. identified the MCPyV [25]. Ever since, epidemiological studies using serological tests have estimated that 60% to 80% of the population is infected with MCPyV [57,58]. Interestingly, the majority of MCC cases (approximately 75%) are linked to MCPyV infection [59–61]. Even though p53 is considered to be a hallmark for NMSCs, Sihto et al. demonstrated that the upregulation of p53 is not a mandatory step for Merkel cell carcinogenesis. In fact, they found p53 to be overexpressed only in 7% of the MCPyV-positive MCC samples suggesting that MCPyV-associated carcinogenesis does not rely on the p53 pathway [62]. Based on the current literature, the proposed mode of MCPyV-induced carcinogenesis relies on at least two critical steps; integration of viral DNA into the cells' genome and loss of its ability to replicate due to accumulated mutations. Following these two steps, the virus produces two main carcinogenic proteins; large T-antigen (LTAg) and small t-antigen (STAg) [62–65]. It has been shown that LTAg specifically binds to tumor suppressor proteins, including p53 (TP53) and members of the Rb family (RB1, RBL1, and RBL2) [66–68].

3. Current Molecular Biomarkers for NMSC

3.1. Telomere Length (TL)

Telomeres are repetitive nucleotide sequences (5'-TTAGGG-3') added on the ends of eukaryotic chromosomes by an enzyme, the telomerase. Combined with specific proteins, telomeres form complexes guarding chromosomic ends from degradation induced by repetitive cell divisions [69] and oxidative stress [70] (Figure 1). Telomerase is an enzyme complex consisting of the catalytic subunit, the human telomerase reverse transcriptase (hTERT) and an RNA template-hTR (human telomere RNA), the telomerase RNA component (TERC), which serves as a template for directing the appropriate telomeric sequences onto the 3' end of a telomeric primer [71]. Given the well-established knowledge that shorter telomeres contribute to cellular senescence [72], both tTL and telomerase activity (TA) have been the subject of research on cancer-related biomarkers. In fact, an increasing body of evidence supports the potential of both serving as diagnostic and prognostic biomarkers for various cancers [73,74]. The underlying hypothesis is that when cellular senescence is combined with excessive environmental burden (for instance UV exposure), the cell may be led to apoptosis. Thus, in theory, it would be reasonable to expect neoplastic cells to possess longer telomeres. On the contrary though, shorter telomeres would render cellular DNA prone to mutations due to replication errors, leading to chromosomal instability and subsequent chromosomal aberrations and therefore, cancer [75]. Nonetheless, from what has been found, it seems that both scenarios may be true for the pathogenesis of NMSC [76], which could be the reason why such a great heterogeneity has been found in association studies [77].

Using Q-FISH for the determination of TL in neoplastic epidermal cells, Yamada-Hishida et al. found that TL was decreased in BD and AK (both had relatively close TL) in relation to BCC and SCC, suggesting that TL estimation in NMSC reflects its biological behavior, such as the metastatic and invasive potential. Moreover, the authors suggested that SCC precursor lesions exhibit a different TL from those of SCC [78]. On the contrary, Wainwright et al. examined BCC and TL in relation to normal skin and reported that telomeres from BCC samples had a variable range of TL (out of the 20 samples they examined 13, had an increased mean TL, while 7 had a shorter TL) [79]. A possible explanation for this variability may be the sampling variability. In other words, the fact that when testing TL from neoplastic cells, one has to bear in mind that cells at one point will differ from those at another despite their relative distance. A solution to this problem was indicated by Han et al., who presented that TL in peripheral blood lymphocytes (PBLs) can be indicative of the skin neoplastic burden and can thus be used as a biomarker. Of note, they found that there was no clear association between TL and the risk of SCC development. By contrast, a shorter telomere length was shown to be associated with an increased risk of BCC [80]. Another study supporting these results was published by Anic et al., who evaluated the relative risk of NMSC development in relation to TL in PBLs. They found that longer telomeres were negatively-associated with BCC and SCC formation (particularly in males), regardless of age [81]. In contrast to the above-mentioned studies, Liang et al. In an equally large series of NMSC cases, reported that there was no association between TL in PBLs and the risk of developing NMSC [82].

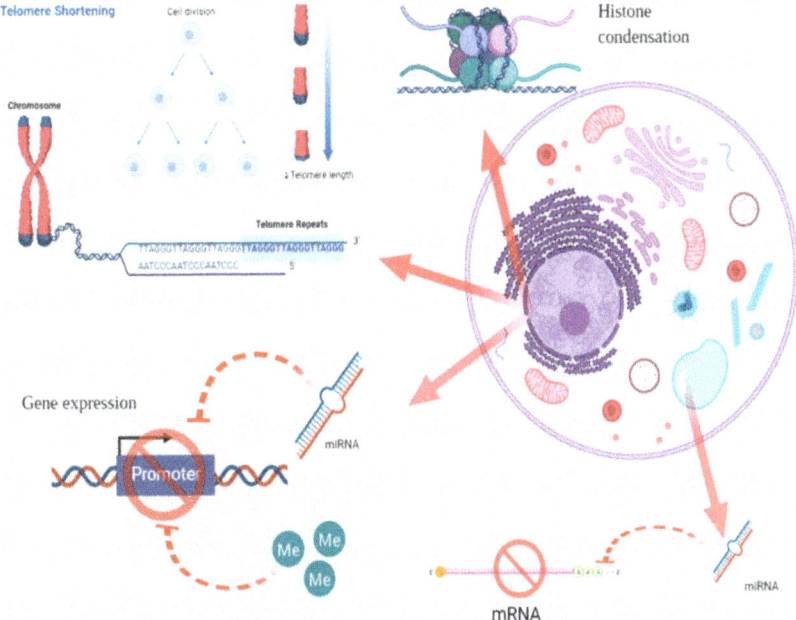

Figure 1. Graphic representation of the underlying pathophysiology of NMSC formation. Carcinogenic mechanisms located in the nuclear apartment involve telomere shortening, histone condensation, inactivation of tumor-suppressor promoters by miRNA and/or methylation. Carcinogenic mechanisms located in the cytosol involve inactivation of mRNAs by miRNAs. Me: methylation.

A rather interesting finding reporting the potential use of TL as a promising indicator of the underlying genetic background giving rise to SCC and the rest of the NMSC was published in the study by Leukfe et al. In detail, they presented that TL distribution is able to differentiate between two types of genetically distinct skin SCCs. The first type exhibits a short/homogeneous TL profile, while the other one a long/heterogeneous TL profile. According to the authors, these findings point out the possibility of two co-existing carcinogenic mechanisms. The first scenario suggests an epidermal stem cell that from some point exhibited accelerated telomere loss which was then passed to his daughter-cells. On the contrary, in the second scenario, which may be the case for the majority of skin SCC cases, a multifocal carcinogenic process occurs with variable proliferation rates at each site, which in hand give rise to variable TLs. In addition, this scenario may explain the profound genetic heterogeneity seen among cancer cells even from the same lesion [76]. This is also important for the determination of the high-risk precursor lesions whose TL resembles that of SCC. Recognizing such lesions would be important for the application of closer monitoring protocols, given that they are more likely to metastasize or recur.

3.2. Telomerase Activity (TA)

As mentioned above, telomerase is composed of two subunits: The catalytic subunit named human telomerase reverse transcriptase (hTERT) and the telomerase RNA component (TERC) for the de novo synthesis of telomeric DNA sequences. The TERT gene, located on the chromosomal area 5p15.33, is the primary regulator of TA via its core promoter region and numerous binding sites which all together serve as transcription regulators. In fact, the main regulatory checkpoint of TA is at its transcription [44]. However, following the genes' pathway upstream, it can be seen that TERT expression is regulated by a number of transcription factors, including c-Myc, Mad1, estrogen receptor, progesterone receptor, AP-1, NF-kB, Rb/E2F factors, CEBP-alpha, and CEBP-beta [83,84] with the Wnt/beta-catenin pathway and the KLF4 being promising candidates as well [85,86]. Of note, it has been shown that TA decreases at the late stages of in utero life, while during ex utero life, it is almost diminished, namely in adult somatic cells [84]. However, an increasing body of evidence supports the notion that most types of cancer cells, among which are skin cancer cells, exhibit an increased TA, mainly due to TERT promoter mutations [87]. Surprisingly, it has also been described that mutations of the TERT gene are of paramount importance for cancer cells derived from tissues with low rates of cellular regeneration [88]. Studying the various TA profiles in skin cancer, Parris I reported that patients with skin cancer exhibited a higher TA than the healthy controls, regardless of the type of cancer. Moreover, a difference in TA was witnessed between the various subtypes of NMSC. In detail, TA was increased in the majority of BD, AK and BCC cases, whereas only in a small number of SCC patients (25%, 3/12). Another interesting finding was the gradual increase in TA in pre-cancerous lesions (42% of AK and BD cases, 11/26) to confirmed cancers (77% of the BCC patients, 10/13) [89]. On the contrary, Boldrini et al. examined a small series of SCCs and BCCs found that SCCs exhibited a higher TA than BCCs, while a close association between hTERT expression and TA was also found. That is of utmost importance, given the relative simplicity of RT-PCR in contrast to TRAP-ELISA, which is the test mostly used for the determination of TA [90]. In a series of 66 patients with NMSC (32 with BCC and 34 with SCC), Griewank et al. found that approximately 50% of both groups had TERT promoter mutations accompanied by significant UV damage in their DNA, with no statistically significant association found with clinicopathologic parameters [91]. In accordance with these findings, Scott et al. reported that TERT promoter mutations were present in 18/23 sporadic BCCs (78%), 13/19 BCCs with nevoid basal cell carcinoma syndrome (68%), 13/26 SCCs (50%), and 1/11 BDs (9%) from a total of 18, 4, 19, and 11 patients, respectively, while being absent in their control group [92]. A finding that has to be noted is that each lesion bears its own genetic fingerprint. That is of utmost importance in cases with multiple lesions where an error in a sampling test should be avoided.

3.3. Epigenetic Modifications

Eukaryotic cells may be subsequent to heritable and non-heritable genomic alterations. Heritable genomic alterations that are not produced by changes in the genomic DNA sequence are summarized as epigenetics [93]. Epigenetic modifications include DNA methylation of the C-5 position of the cytosine ring within the promoter's CpG island, histone methylation and acetylation, and miRNA-mediated gene regulations. Separately and combined, these alterations regulate the chromatin formation and packaging and thus regulate gene transcription by modifying their accessibility [94]. It is accepted that epigenetic modifications reflect the environmental burden of an organism through its exposure to various toxicants and carcinogens [95].

3.3.1. CpG Island Methylation (CIM)

DNA methylation is one the most important regulatory mechanisms for gene expression. In normal cells, it assures the proper regulation of gene expression and stable gene silencing. This is achieved through the recruitment of DNA methyltransferases (DNMTs) in order to introduce methyl groups in cytosine within CpG dinucleotides by creating covalent bonds between them. In fact, CpG dinucleotides may appear in large clusters known as CpG islands (Figure 1). Intense research in cancer biology has identified global genomic hypomethylation as one of the leading factors for genomic instability and oncogene activation, whereas a number of tumor suppressor genes are silenced due to hypermethylated CpG islands [96], while global hypomethylation of lamina-associated domains (LAD) may be another aspect of the deregulated methylome [97]. In cutaneous melanoma, it was demonstrated that promoter hypomethylation and intragenic hypermethylation of specific genes are associated with tumor aggressiveness due to the alteration of extracellular matrix components and the upregulation of matrix metalloproteinases [98–100]. This highlights the clinical potential of deregulated methylation status as a hallmark for carcinogenesis, allowing the recognition of various methylation patterns as biomarkers for diagnosis and prognosis [101] (Table 1). Methylation studies focusing on cSCC, have demonstrated various patterns. For instance, numerous promoters have been found to be hypermethylated, among which are the cell cycle regulator CDKN2A [102], cadherin CDH1 [103,104] and CDH13 [105], transcription factor FOXE1 [106], modulators of Wnt signaling SFRPs [107] and FRZB [108], positive regulators of apoptosis ASC [109], G0S2 [110], DAPK1 [111], and miRNA-204 [112], as well as the hypomethylation of the DSS1 gene [113]. Hervás-Marín et al. compared low-risk and high-risk SCC and succeeded in identifying specific modifications of the methylation status using genome-wide DNA methylation profiling. In detail, they demonstrated a differential methylation status between the two pathological stages, with low-risk SCCs being hypomethylated and high-risk SCCs hypermethylated. According to the authors, this finding may suggest a sequential approach of SCC formation, where UV-exposure leads to hypomethylation and thus foretells the premalignant and low-risk stages of cSCC, while advanced stages of SCC present a hypermethylated status [101]. As regards the evaluation of the methylation status of BCC, Goldberg et al. presented the FHIT promoter to be hypomethylated [114], while Heitzer et al. found the hypermethylated PTCH promoter only in a small number of cases [115]. Darr et al. examined metastatic BCCs and SCCs compared to their non-metastatic counterparts. They found that both metastatic entities exhibited a differential methylation status from the non-metastatic ones with pronounced global hypomethylation, as well as at tumor suppressor genes and PRC2 target genes. Moreover, MYCL2 was specifically found to be demethylated in metastatic cases. Of note, the authors highlighted the resemblance between the methylation pattern of metastatic BCC and cSCC regardless of the metastatic capacity [108]. Greenberg et al., studying a series of MCCs, demonstrated that the tumor suppressor p14-ARK was hypermethylated [116]. Moreover, hypermethylated promoters have also been found in DUSP2, CDKN2A, and members of the RASSF family [117]. The concomitant analysis of overexpressed proteins derived from methylated genes and hallmark mutations of skin cancers through high-sensitive molecular techniques is representing a promising strategy for the early diagnosis of tumors and to define the prognosis of patients [118].

Table 1. NMSC-related genomic loci, their methylation status, and their effect on cellular level.

Gene Target	Methylation Status	Type of NMSC	Cellular Effect	Reference
CDKN2A	Hypermethylated	SCC	Cell cycle deregulation	Brown et al. [102]
CDH1	Hypermethylated	SCC	Cellular environment deregulation	Chiles et al. [103] Murao et al. [104]
CDH13	Hypermethylated	SCC	Cellular environment deregulation	Takeuchi et al. [105]
FOXE1	Hypermethylated	SCC	Modulator of Wnt signaling	Venza et al. [106]
SFRPs	Hypermethylated	SCC	Modulator of Wnt signaling	Liang et al. [107]
FRZB	Hypermethylated	SCC	Modulator of Wnt signaling	Darr et al. [108]
ASC	Hypermethylated	SCC	Deregulation of apoptosis	Meier et al. [109]
G0S2	Hypermethylated	SCC	Deregulation of apoptosis	Nobeyama et al. [110]
DAPK1	Hypermethylated	SCC	Deregulation of apoptosis	Li et al. [111]

Table 1. *Cont.*

Gene Target	Methylation Status	Type of NMSC	Cellular Effect	Reference
miRNA-204	Hypermethylated	SCC	Deregulation of apoptosis	Toll et al. [112]
DSS1	Hypomethylation	SCC	Deregulated post-translational protein modification	Venza et al. [113]
Global DNA	Hypomethylation	SCC (benign)	Restricted genomic silencing	Hervás-Marín et al. [101]
Global DNA	Hypermethylation	SCC (aggressive)	Extensive genomic silencing	Hervás-Marín et al. [101]
FHIT promoter	Hypomethylated	BCC	Replication stress and DNA damage	Goldberg et al. [114]
PTCH promoter	Hypermethylated	BCC (small number of cases)	Deactivation of tumor suppressor genes	Heitzer et al. [115]
MYCL2	Hypomethylated	BCC (metastatic)	Activation of proto-oncogene	Darr et al. [108]
p14-ARK	Hypermethylated	MCC	Deactivation of tumor suppressor genes	Greenberg et al. [116]
DUSP2, CDKN2A promoter	Hypermethylated	MCC	Deactivation of tumor suppressor genes	Harms et al. [117]

3.3.2. Histone Methylation and Acetylation

Histones are a family of five basic proteins (H1/H5, H2A, H2B, H3, and H4) whose role is to react with DNA strands in the nucleus assisting its dense packaging into chromatin and chromosomes. Histones H2A, H2B, H3, and H4 form a reel of dimers (the octameric nucleosome core) around which DNA is wrapped, while histones H1/H5 link nucleosomes together, allowing for an even higher degree of density (Figure 1). A key feature of histones is the presence of the N-terminal tail regions, which are rich in lysine residues. The histone tails can undergo extensive modifications, including methylation, acetylation, phosphorylation, sumoylation, and uquitinylation [119,120]. However, acetylation and methylation are the most well-studied aspects of histone modification, particularly in the setting of cancer. The acetylation and deacetylation of lysine residues modifies the net positive charge (decreasing or increasing it accordingly). Furthermore, the introduction of acetyl-groups induces a decreased affinity between histones and DNA, allowing for various transcription factors to reach regulatory areas such as gene promoters, while deacetylation has the opposite effect on gene expression by increasing the affinity between DNA and the histone complex [116]. Histone acetylation and deacetylation are catalyzed by the specific enzymes, histone acetyltransferases (HATs) and histone deacetyltransferases (HDACs), respectively. Histones are mainly methylated on the lysine and arginine residues of H3 and H4 tails [93]. The introduction of methyl-groups increases the hydrophobicity of histone proteins, inducing their tighter packing and thus inhibiting DNA transcription. Notably, it has been described that the restoration of normal histone density (reduction of DNA methylation and increase of histone acetylation) allows for the reactivation of the silenced tumor suppressor genes Cip1/p21 and p16 [121]. Rao et al. investigated the activation status of EZH2 (a histone methyltransferase of the polycomb repressive complex 2) and its related proteins in the context of aggressive BCCs. EZH2 is closely associated with the Sonic Hedgehog pathway [122]. According to their findings, EZH2 was upregulated (as in other studies [123]), allowing for a stratification between pathological stages. On the contrary, upregulated H3K27me3 and 5hmC were positively associated with a more benign phenotype. Finally, the authors were able to discriminate BCCs from non-malignant epidermal cells through the upregulated levels NSD2, MOF, H3K27me3, and 5hmC [124]. Harms et al. investigated a series of MCCs and found that EZH2 was deregulated, inducing gene silencing via histone H3 lysine 27 trimethylation and was thus associated with unfavorable characteristics, such as disease progression and a poorer prognosis [117,125]. However, even though histone methylation/acetylation has been extensively investigated in melanoma [116,126], research on NMSCs is limited. Indeed, it was recently demonstrated that the methylation of H3K4 is associated with the neoplastic transformation of melanocytes that evolve into cutaneous melanoma [127]. These results suggest that the epigenetic modification of histones' methylation status could represent a promising epigenetic therapy for melanoma and other tumors [126].

3.3.3. MicroRNAs (miRNAs or miRs)

miRNAs are small single-stranded non-coding RNAs of 18–25 nucleotides length. Their discovery in 1993 from two research groups working on *Caenorhabditis elegans* proved to be a milestone of what is now considered a true breakthrough in molecular biology [128,129]. However, for a number of years, the properties miRNAs remained poorly understood. Surprisingly, miRNA production is a refined, multi-step process, where specific DNA transcripts produce primary miRNAs (pri-miRNAs), which are processed into precursor miRNAs (pre-miRNAs) and then into mature miRNAs. Mature miRNAs have the potential to target specific mRNAs, leading to their degradation or inhibiting their translation into proteins. This is possible either through an interaction with the 3'-untranslated region (3' UTR) of the target mRNA (in which case its expression is inhibited) [130] or through binding with other regions, such as the 5'-untranslated region (5' UTR), coding sequence and gene promoters [131]. Of note however, miRNAs are able to regulate not only protein translation, but also gene expression. In detail, miRNAs have been found to be able to positively regulate gene expression under certain conditions [132]. This is possible as miRNAs are able to move through different cellular compartments [133] (Figure 1). However, miRNAs are not restricted to the cytosol. A number of studies have demonstrated the presence of miRNAs in the extracellular compartment, both in a free state and packed in various carriers, such as high density lipoprotein particles, apoptotic bodies, and others [134]. Indeed, in addition to their small size and hairpin-loop structure, they are unreachable to the various free RNases, allowing them to maintain their structural integrity [135]. Thus, isolating them from a variety of clinical specimens is possible. Lastly, it has been well established that miRNAs are actively secreted by a variety of cancer cells into the circulation [73]. However, each type of cancer expresses different miRNAs; thus, in this manner, each type of cancer creates its own molecular profile. This is of utmost importance when considering miRNAs as biomarkers for monitoring cellular activity and the genomic/proteomic status. Even though miRNAs can be isolated both from tissue samples and from biological fluids (serum, plasma, and urine), circulating miRNAs are the first choice in the clinical setting. This is due to the fact that tissue miRNA sampling is an invasive technic lacking the ability to provide reproducible results regardless of the operator and area of sampling [136]. At present, several studies have identified sets of miRNAs specific for different tumors, including lung cancer, mesothelioma, bladder cancer, colorectal cancer, glioblastoma multiforme, oral cancer, uveal melanoma, hematological malignancies, etc. [137–144].

Regarding NMSCs, owing to the dominance of BCCs among all other tumor types, numerous studies have focused on the identification of potential miRNA markers. Sand et al. used next-generation sequencing of the basal cell carcinoma miRNome and succeeded in identifying a number of upregulated miRNAs, of which the 10 most increased were hsa-miR-223-3p, hsa-miR-197-3p, hsa-miR-342-3p, hsa-miR-505-3p, hsa-miR-204-5p, hsa-miR-941, hsa-miR-145-5p, hsa-miR-301b-3p, hsa-miR-452-5p, and hsa-miR-191-5p [145]. Yi et al. found that miR-203, a specifically expressed miRNA in the epidermis [146], is consistently downregulated in cases of BCC. Moreover, they proved that c-JUN suppressed miR-203, while miR-203 also targeted c-JUN, creating an inhibitory loop. In addition, miR-203 was further suppressed by the synergistic oncogenic activity of the Sonic Hedgehog and EGFR pathways. It is rather interesting that various studies have identified c-JUN as a potent oncogene, mediating its action downstream of the Sonic Hedgehog pathway [147]. Thus, a simultaneous activation of the Sonic Hedgehog and EGFR pathways, in addition to a potential crosstalk between them may result in BCC formation. Given the inhibitory effect of miR-203 c-JUN, researchers have investigated the therapeutic potential of miR-203 administration. Indeed, high levels of miR-203 have been shown to result in a decreased c-JUN and p63 expression, indicating the effective suppression of target genes [148]. Hu et al. examined 86 patients with BCC in order to explore the association between the expression level of miR-34a in serum and the clinical prognosis. According to their findings, patients with BCC exhibited lower miR-34a levels compared to healthy controls. Data analysis further revealed that miR-34a was upregulated in cases with a larger tumor diameter, the absence of lymph node infiltration and non-invasive disease. Moreover, miR-34a was positively associated with various survival parameters, such as median progression-free survival, median overall survival, and the overall survival rate. However, no association was found with pathological staging or the primary site. On the contrary, cases with a profound downregulation of miR-34a presented a poor prognosis [149].

In SCC, numerous miRNAs have been found to be dysregulated. Some of these (namely miR-21, miR-205, miR-365, miR-31, miR-135b, miR-424, miR-320, miR-222, miR-15a, miR-142, and miR-186) have been shown to possess carcinogenic properties by targeting key genetic modulators, such as the PTEN, PDCD4, GRHL3, HOXA9, and RhoBTB genes or the AKT/mTOR pathway [150,151]. There is sufficient evidence to indicate that these genes are involved in crucial carcinogenic steps, such as tumor growth, invasion, migration, the maintenance of stem cell properties and the evasion of apoptosis [151]. On the contrary, there is a wide panel of carcinoprotective miRNAs (miR-20a, miR-203, miR-181a, miR-125b, miR-34a, miR-148a, miR-214, miR-124, miR-204, and miR-199a), which have been found to regulate genes, such as HMGB1, SIRT6, MMPs, MAP kinases, KRAS, LIMK1, c-MYC, SHP2, CD44, BCAM, FZD6, DDR1, and ERKs. The potential action is described to be via the regulation of the cell cycle, epithelial–mesenchymal transition, and stemness, while they have also been found to promote cellular apoptosis and senescence [152]. A number of studies have evaluated the association of various miRNAs with clinocopathological features. miR-205 has exhibited an association with various pathological features of a poor prognosis, such as desmoplasia, perineural invasion and infiltrative patterns, while clinically it has been linked to local recurrence [153,154]. Recently, Gong et al. described that miR-221 also has carcinogenic properties. This is achieved as miR-221 has been found to interact with PTEN, which is a key oncogene. Notably, the authors pinpointed the potential development of anti-miR-221 antibodies, assisting both diagnosis and treatment [155]. On the contrary, miR-203 expression was shown to be associated with a favorable prognosis, as it was primarily found in well-differentiated zones only and rarely in the invasion front [153]. Zhang et al. found that SCC patients with low miR-20a levels exhibited a significantly poorer overall survival than those with a high miR-20a expression. Moreover, miR-20a expression was closely associated with the TNM stage, as it was proven that a low level of miR-20a expression was more frequently exhibited in tumors with an advanced TNM stage [156]. Several studies have also examined the expression profiles of various miRNAs in MCC. Ning et al. used next-generation sequencing of small RNA libraries on tissue samples and identified the MCC miRNome. In total, eight miRNAs were overexpressed (miR-502-3p, miR-9, miR-7, miR-340, miR-182, miR-190b, miR-873, and miR-183) and three miRNAs were suppressed (miR-3170, miR-125b, and miR-374c) in contrast to other forms of NMSCs. In situ hybridization further proved that miR-182 was abundant within cancer cells. The concomitant evaluation of the expression profiles of four miRNAs (miR-182, miR-183, miR-190b, and miR-340) in the MCPyV-negative cell line, MCC13, proved that they were downregulated. Thus, they proposed the possible diagnostic use of this miRNA panel in cases of MCPyV-positive MCC [157]. Veija et al. compared the miRNAome between MCPyV-positive and MCPyV-negative MCCs. According to their findings, miR-30a, miR-34a, miR-142-3p, and miR-1539 were overexpressed (2.5 to 5 times) in MCPyV-positive MCCs, while miR-181d exhibited a 3.5-fold higher expression in MCPyV-negative MCCs [158]. Renwick et al. used miRNA FISH in formalin-fixed paraffin-embedded tissues and succeeded in correctly discerning BCC from MCC, based on the overexpression of miR-205 and miR-375, respectively [159]. An important finding also derived from the study by Moens et al. who evaluated the secretion of various miRNAs in exosomes using RT-PCR. They succeeded in identifying the presence of miR-30a, miR-125b, miR-183, miR-190b, and miR-375 in exosomes [160]. This finding highlights the clinical potential of circulating miRNAs as biomarkers for MCC. In this context, the analysis of circulating tumor DNA and circulating miRNAs has been translated into clinical practice to predict the clinical-pathological features of tumors thus ameliorating the diagnostic and therapeutic strategies available for cancer patients [161]. However, solid evidence for the clinical relevance of extracellular miRNAs is still lacking. Moreover, despite the fact that numerous studies emerge daily, enriching the MCC-related miRNA panel, only a few of these have compared the expression profiles between malignant and non-malignant Merkel cells, and even fewer have tested the clinical or pathological relevance of these profiles [160,162]. miRNAs are recognized also as important biomarkers for the management of cutaneous melanoma. In this context, several studies have identified sets of miRNAs strictly associated to the development and progression of melanoma. In particular, Tao and colleagues (2019) have identified five miRNAs (miR-25, miR-204, miR-211, miR-510, and miR-513c) associated with survival of melanoma patients [163]. In the same manner, Hanniford et al. have identified a 4-miRNA signature (miR-150–5p, miR-15b-5p, miR-16–5p, and miR-374b-3p) predictive for the development of melanoma brain metastases [164]. Notably, some miRNAs, in particular the miR-510, are associated with both melanoma clinical features. A summary of the various miRNA expressions patterns and their clinical significance for NMSCs is presented in Table 2.

Table 2. Deregulated microRNA expression profiles and their clinical relevance for NMSC.

miRNA	Expression Status	Type of NMSC	Possible Significance	Reference
hsa-miR-223-3p, hsa-miR-197-3p, hsa-miR-342-3p, hsa-miR-505-3p, hsa-miR-204-5p, hsa-miR-941, hsa-miR-145-5p, hsa-miR-301b-3p, hsa-miR-452-5p, hsa-miR-191-5p,	Upregulated	BCC	Diagnosis	Sand et al. [145]
miR203	Downregulated	BCC	Diagnosis, Therapy	Yi et al. [146]
miR-34a	Downregulated	BCC	Prognosis	Hu et al. [149]
miR-21, miR-205, miR-365, miR-31, miR-135b, miR-424, miR-320, miR-222, miR-15a, miR-142, miR-186	Upregulated	SCC	Diagnosis	Mizrahi et al. [150], Yu et al. [151]
miR-20a, miR-203, miR-181a, miR-125b, miR-34a, miR-148a, miR-214, miR-124, miR-204, miR-199a	Downregulated	SCC	Diagnosis	García-Sancha et al. [152]
miR-205	Upregulated	SCC	Diagnosis, Prognosis	Cañueto et al. [153], Stojadinovic et al. [154]
miR-221	Upregulated	SCC	Diagnosis, therapy	Gong et al. [155]
miR-203	varied	SCC	Prognosis	Cañueto et al. [153]
miR-20a	Varied	SCC	Prognosis	Zhang et al. [156]

Table 2. Cont.

miRNA	Expression Status	Type of NMSC	Possible Significance	Reference
miR-502-3p, miR-9, miR-7, miR-340, miR-182, miR-190b, miR-873, miR-183	Upregulated	MCC	Diagnosis	Ning et al. [157]
miR-3170, miR-125b, miR-374c	Downregulated			
miR-182, miR-183, miR-190b, miR-340	Downregulated in MCPyV-negative cell line			
miR-30a, miR-190b, miR-142-3p, miR-1539	Upregulated	MCPyV-positive MCCs	Diagnosis	Veija et al. [158]
miR-181d		MCPyV-negative MCCs		
miR-375	Upregulated	MCC	Diagnosis	Renwick et al. [159]
miR-30a, miR-125b, miR-183, miR-190b, miR-375	Upregulated	MCC	Diagnosis	Moens group [160]

4. Biomarkers under Evaluation

Micronuclei Frequency (MNf)

Micronuclei (MN), or Howell–Jolly bodies, are small cytoplasmic formations unsheathed in a nuclear envelope. In nature, they represent acentric chromatid/chromosome fragments (as a result of DNA damage) or whole chromatids/chromosomes (due to mitotic spindle failure, kinetochore damage, centromeric DNA hypomethylation and defects in the cell cycle control system) that are not included in the nucleus during telophase. Instead, they form small DNA-containing structures that are just a fraction of the size of the nucleus [165,166]. A large number of studies have indicated the promising potential of MN frequency (MNf) as a biomarker for diagnostic, prognostic and predictive use in various types of cancer, among which are those of the lung, bladder, and colorectal cancer [167,168]. However, both melanoma and NMSC have not been extensively studied with regards to their MNf status. Nonetheless, there is evidence that in premalignant cell lines (for example keratinocytes), MNf is higher than in normal skin lines [169], while chromosomal aberrations due to UVA and UVB skin exposure also result in an increased MNf [170,171]. Taking all these findings into consideration, it can be hypothesized that MNf as part of a wider panel of biomarkers, can be used not only for the diagnosis of NMSC, but also for a close and convenient monitoring for the early detection of tumor regression or progression.

5. Conclusions

NMSC is the most common type of cancer worldwide, representing an immense burden for both patients and healthcare systems. However, if diagnosed in an early stage, a great number of these cases will probably have a definitive care. Moreover, the vast majority of NMSC cases have well-studied causative factors, allowing for the establishment of screening protocols meant for high-risk groups. On the contrary, it is suggested that the macroscopic examination of the skin largely fails to assist secondary prevention improvement. Thus, the introduction of more sensitive and specific modes of diagnosis is required. The present review aimed to systematically suggest that molecular biomarkers are able to achieve this goal. In fact, molecular biomarkers seem to be promising candidates, not only for early detection, but also for the achievement of the corner stone of effective care which is personalized medicine. Despite the fact that NMSCs are distinct entities, they have been proven to share some common features to a certain extent. The hypermethylated E-cadherin (CDH1) promoter and the deregulated expression profile of miR-203 are some of the BCC/SCC shared biomarkers. However, as presented above, even if current literature suggests the possible clinical significance of various molecular targets (micronuclei frequency, extracellular miRNAs, histone methylation/acetylation) solid evidence on this topic is still missing. This highlights the need for further validation first through in vivo and then through large cohort studies where panels of sensitive and specific biomarkers will be evaluated both for their ability to detect and for their availability to foretell the prognosis. Unfortunately, a great disadvantage of NMSC biomarkers is the inability to specifically locate a lesion that has not made itself clinically/macroscopically evident yet. Thus, research for biomarkers has to create panels that will be not only disease-sensitive/specific but also site-sensitive/specific and therefore being able to discern between different body regions or between skin and mucous membrane cancers.

Author Contributions: T.K.N. was involved in the conceptualization of the study and in the investigative aspects of the study as well as in formal analysis and writing of the original draft. L.F. was involved in the investigative aspects of the study and in formal analysis, and the writing of the manuscript. K.L., S.K.-K., M.S., and A.K. were involved in the writing, reviewing, and editing of the manuscript. E.C., D.A.S., and A.T. were involved in the reviewing and editing of the manuscript. J.T. was involved in the conceptualization and methodology of the study, and in the writing, reviewing, and editing of the manuscript, as well as in study supervision. All authors have read and agreed to the published version of the manuscript.

Funding: No external funding was involved in this review.

Conflicts of Interest: The authors declare no conflict of interest.

Abbreviations

NMSC	non-melanoma skin cancer
BCC	basal cell carcinoma
SCC	squamous cell carcinoma
MCC	Merkel cell carcinoma
BD	Bowen's Disease
AK	Actinic Keratosis
UV	Ultraviolet
β-HPV	β-Human papilloma virus
UVB	ultraviolet B
UVA	Ultraviolet A
EGFR	epidermal growth factor receptor
FGFR	fibroblast growth factor receptors
EBV	Epstein–Barr virus
MCPyV	Merkel Cell Polyomavirus
LTAg	large T-antigen
STAg	small t-antigen
TL	Telomere length
hTERT	human telomerase reverse transcriptase
hTR	human telomere RNA
TERC	telomerase RNA component
TA	telomerase activity
PBL	peripheral blood lymphocytes
CIM	CpG island methylation
DNMTs	DNA methyltransferases
LAD	lamina-associated domains
HATs	acetyltransferases
HDACs	histone deacetyltransferases
pri-miRNAs	primary miRNAs
pre-miRNAs	precursor miRNAs

3' UTR	3'-untranslated region
5' UTR	5'-untranslated region
MN	Micronuclei
MNf	Micronuclei frequency

References

1. Leiter, U.; Eigentler, T.; Garbe, C. Epidemiology of skin cancer. In *Advances in Experimental Medicine and Biology*; Springer: New York, NY, USA, 2014; Volume 810, pp. 120–140.
2. Rigel, D.S.; Friedman, R.J.; Kopf, A.W. Lifetime risk for development of skin cancer in the U.S. population: Current estimate is now 1 in 5. *J. Am. Acad. Dermatol.* **1996**, *35*, 1012–1013. [CrossRef]
3. Apalla, Z.; Nashan, D.; Weller, R.B.; Castellsagué, X. Skin Cancer: Epidemiology, Disease Burden, Pathophysiology, Diagnosis, and Therapeutic Approaches. *Dermatol. Ther.* **2017**, *7*, 5–19. [CrossRef] [PubMed]
4. Lomas, A.; Leonardi-Bee, J.; Bath-Hextall, F. A systematic review of worldwide incidence of nonmelanoma skin cancer. *Br. J. Dermatol.* **2012**, *166*, 1069–1080. [CrossRef] [PubMed]
5. Smith, H.; Wernham, A.; Patel, A. When to suspect a non-melanoma skin cancer. *BMJ* **2020**, *368*, m692. [CrossRef] [PubMed]
6. Samarasinghe, V.; Madan, V. Nonmelanoma skin cancer. *J. Cutan. Aesthet. Surg.* **2012**, *5*, 3. [CrossRef] [PubMed]
7. Grabowski, J.; Saltzstein, S.L.; Sadler, G.R.; Tahir, Z.; Blair, S. A comparison of Merkel cell carcinoma and melanoma: Results from the California Cancer Registry. *Clin. Med. Oncol.* **2008**, *2*, 327–333. [CrossRef]
8. Yanofsky, V.R.; Mercer, S.E.; Phelps, R.G. Histopathological Variants of Cutaneous Squamous Cell Carcinoma: A Review. *J. Skin Cancer* **2011**, *2011*, 210813. [CrossRef]
9. Kim, R.H.; Armstrong, A.W. Nonmelanoma Skin Cancer. *Dermatol. Clin.* **2012**, *30*, 125–139. [CrossRef]
10. Wong, C.S.M. Basal cell carcinoma. *BMJ* **2003**, *327*, 794–798. [CrossRef]
11. Dacosta Byfield, S.; Chen, D.; Yim, Y.M.; Reyes, C. Age distribution of patients with advanced non-melanoma skin cancer in the United States. *Arch. Dermatol. Res.* **2013**, *305*, 845–850. [CrossRef]
12. Peterson, S.C.; Eberl, M.; Vagnozzi, A.N.; Belkadi, A.; Veniaminova, N.A.; Verhaegen, M.E.; Bichakjian, C.K.; Ward, N.L.; Dlugosz, A.A.; Wong, S.Y. Basal cell carcinoma preferentially arises from stem cells within hair follicle and mechanosensory niches. *Cell Stem Cell* **2015**, *16*, 400–412. [CrossRef] [PubMed]
13. Kwasniak, L.A.; Garcia-Zuazaga, J. Basal cell carcinoma: Evidence-based medicine and review of treatment modalities. *Int. J. Dermatol.* **2011**, *50*, 645–658. [CrossRef] [PubMed]
14. Rodriguez-Vigil, T.; Vázquez-López, F.; Perez-Oliva, N. Recurrence rates of primary basal cell carcinoma in facial risk areas treated with curettage and electrodesiccation. *J. Am. Acad. Dermatol.* **2007**, *56*, 91–95. [CrossRef]
15. Habif, T.P. *Clinical Dermatology: A Color Guide to Diagnosis and Therapy*; Elsevier: St. Louis, MO, USA, 2016; ISBN 9780323266079.
16. Yan, W.; Wistuba, I.I.; Emmert-Buck, M.R.; Erickson, H.S. Squamous Cell Carcinoma—Similarities and Differences among Anatomical Sites. *Am. J. Cancer Res.* **2011**, *1*, 275–300. [PubMed]
17. Paulson, K.G.; Park, S.Y.; Vandeven, N.A.; Lachance, K.; Thomas, H.; Chapuis, A.G.; Harms, K.L.; Thompson, J.A.; Bhatia, S.; Stang, A.; et al. Merkel cell carcinoma: Current US incidence and projected increases based on changing demographics. *J. Am. Acad. Dermatol.* **2018**, *78*, 457–463. [CrossRef] [PubMed]
18. Schadendorf, D.; Lebbé, C.; Zur Hausen, A.; Avril, M.-F.; Hariharan, S.; Bharmal, M.; Rgen, J.; Becker, C.; Bharmal, M.; De, J.B.-H.; et al. Merkel cell carcinoma: Epidemiology, prognosis, therapy and unmet medical needs. *Eur. J. Cancer* **2017**, *71*, 53–69. [CrossRef] [PubMed]
19. Lunder, E.J.; Stern, R.S. Merkel-cell carcinomas in patients treated with methoxsalen and ultraviolet a radiation. *N. Engl. J. Med.* **1998**, *339*, 1247–1248. [CrossRef]
20. Aw, D.; Silva, A.B.; Palmer, D.B. Immunosenescence: Emerging challenges for an ageing population. *Immunology* **2007**, *120*, 435–446. [CrossRef]
21. Engels, E.A.; Frisch, M.; Goedert, J.J.; Biggar, R.J.; Miller, R.W. Merkel cell carcinoma and HIV infection. *Lancet* **2002**, *359*, 497–498. [CrossRef]
22. Toker, C. Trabecular carcinoma of the skin. *Arch. Dermatol.* **1972**, *105*, 107–110. [CrossRef]
23. Coggshall, K.; Tello, T.L.; North, J.P.; Yu, S.S. Merkel cell carcinoma: An update and review: Pathogenesis, diagnosis, and staging. *J. Am. Acad. Dermatol.* **2018**, *78*, 433–442. [CrossRef] [PubMed]

24. Duprat, J.P.; Landman, G.; Salvajoli, J.V.; Brechtbühl, E.R. A review of the epidemiology and treatment of Merkel cell carcinoma. *Clinics* **2011**, *66*, 1817–1823. [CrossRef] [PubMed]
25. Feng, H.; Shuda, M.; Chang, Y.; Moore, P.S. Clonal integration of a polyomavirus in human Merkel cell carcinoma. *Science* **2008**, *319*, 1096–1100. [CrossRef] [PubMed]
26. Clark, C.M.; Furniss, M.; Mackay-Wiggan, J.M. Basal cell carcinoma: An evidence-based treatment update. *Am. J. Clin. Dermatol.* **2014**, *15*, 197–216. [CrossRef]
27. Bromfield, G.; Dale, D.; De, P.; Newman, K.; Rahal, R.; Shaw, A. *Canadian Cancer Statistics 2015 Special Topic: Predictions of the Future Burden of Cancer in Canada*; Canadian Cancer Society, Government of Canada: Toronto, ON, Canada, 2015.
28. Non-Melanoma Skin Cancer Mortality Statistics | Cancer Research UK. Available online: https://www.cancerresearchuk.org/health-professional/cancer-statistics/statistics-by-cancer-type/non-melanoma-skin-cancer/mortality#heading-Two (accessed on 21 March 2020).
29. Chren, M.M.; Torres, J.S.; Stuart, S.E.; Bertenthal, D.; Labrador, R.J.; Boscardin, W.J. Recurrence after treatment of nonmelanoma skin cancer: A prospective cohort study. *Arch. Dermatol.* **2011**, *147*, 540–546. [CrossRef]
30. Chren, M.M.; Linos, E.; Torres, J.S.; Stuart, S.E.; Parvataneni, R.; Boscardin, W.J. Tumor recurrence 5 years after treatment of cutaneous basal cell carcinoma and squamous cell carcinoma. *J. Investig. Dermatol.* **2013**, *133*, 1188–1196. [CrossRef]
31. Kaldor, J.; Shugg, D.; Young, B.; Dwyer, T.; Wang, Y.G. Non-melanoma skin cancer: Ten years of cancer-registry-based surveillance. *Int. J. Cancer* **1993**, *53*, 886–891. [CrossRef]
32. Didona, D.; Paolino, G.; Bottoni, U.; Cantisani, C. Non melanoma skin cancer pathogenesis overview. *Biomedicines* **2018**, *6*, 6. [CrossRef]
33. Silverberg, M.J.; Leyden, W.; Warton, E.M.; Quesenberry, C.P.; Engels, E.A.; Asgari, M.M. HIV infection status, immunodeficiency, and the incidence of non-melanoma skin cancer. *J. Natl. Cancer Inst.* **2013**, *105*, 350–360. [CrossRef]
34. Chew, M.H.; Yeh, Y.T.; Toh, E.L.; Sumarli, S.A.; Chew, G.K.; Lee, L.S.; Tan, M.H.; Hennedige, T.P.; Ng, S.Y.; Lee, S.K.; et al. Critical evaluation of contemporary management in a new Pelvic Exenteration Unit: The first 25 consecutive cases. *World J. Gastrointest. Oncol.* **2017**, *9*, 218–227. [CrossRef]
35. Nikolaou, V.; Stratigos, A.J.; Tsao, H. Hereditary Nonmelanoma Skin Cancer. *Semin. Cutan. Med. Surg.* **2012**, *31*, 204–210. [CrossRef] [PubMed]
36. Samarasinghe, V.; Madan, V.; Lear, J.T. Focus on Basal cell carcinoma. *J. Skin Cancer* **2011**, *2011*, 328615. [CrossRef] [PubMed]
37. Smeets, N.W.J.; Kuijpers, D.I.M.; Nelemans, P.; Ostertag, J.U.; Verhaegh, M.E.J.M.; Krekels, G.A.M.; Neumann, H.A.M. Mohs' micrographic surgery for treatment of basal cell carcinoma of the face—Results of a retrospective study and review of the literature. *Br. J. Dermatol.* **2004**, *151*, 141–147. [CrossRef] [PubMed]
38. Welsh, M.M.; Karagas, M.R.; Kuriger, J.K.; Houseman, A.; Spencer, S.K.; Perry, A.E.; Nelson, H.H. Genetic Determinants of UV-Susceptibility in Non-Melanoma Skin Cancer. *PLoS ONE* **2011**, *6*, e20019. [CrossRef]
39. Saijo, S.; Kodari, E.; Kripke, M.L.; Strickland, F.M. UVB irradiation decreases the magnitude of the Th1 response to hapten but does not increase the Th2 response. *Photodermatol. Photoimmunol. Photomed.* **1996**, *12*, 145–153. [CrossRef]
40. Morison, W.L.; Bucana, C.; Kripke, M.L. Systemic suppression of contact hypersensitivity by UVB radiation is unrelated to the UVB-induced alterations in the morphology and number of Langerhans cells. *Immunology* **1984**, *52*, 299–306.
41. McGillis, S.T.; Fein, H. Topical treatment strategies for non-melanoma skin cancer and precursor lesions. *Semin. Cutan. Med. Surg.* **2004**, *23*, 174–183. [CrossRef]
42. Christensen, S.R. Recent advances in field cancerization and management of multiple cutaneous squamous cell carcinomas [version 1; referees: 2 approved]. *F1000Research* **2018**, *7*. [CrossRef]
43. Reifenberger, J.; Wolter, M.; Knobbe, C.B.; Köhler, B.; Schönicke, A.; Scharwächter, C.; Kumar, K.; Blaschke, B.; Ruzicka, T.; Reifenberger, G. Somatic mutations in the PTCH, SMOH, SUFUH and TP53 genes in sporadic basal cell carcinomas. *Br. J. Dermatol.* **2005**, *152*, 43–51. [CrossRef]
44. Pellegrini, C.; Maturo, M.G.; Di Nardo, L.; Ciciarelli, V.; Gutiérrez García-Rodrigo, C.; Fargnoli, M.C. Understanding the molecular genetics of basal cell carcinoma. *Int. J. Mol. Sci.* **2017**, *18*, 2485. [CrossRef]
45. Dotto, G.P.; Rustgi, A.K. Squamous Cell Cancers: A Unified Perspective on Biology and Genetics. *Cancer Cell* **2016**, *29*, 622–637. [CrossRef] [PubMed]

46. Lawrence, M.S.; Sougnez, C.; Lichtenstein, L.; Cibulskis, K.; Lander, E.; Gabriel, S.B.; Getz, G.; Ally, A.; Balasundaram, M.; Birol, I.; et al. Comprehensive genomic characterization of head and neck squamous cell carcinomas. *Nature* **2015**, *517*, 576–582.
47. Freed-Pastor, W.A.; Prives, C. Mutant p53: One name, many proteins. *Genes Dev.* **2012**, *26*, 1268–1286. [CrossRef] [PubMed]
48. Muller, P.A.J.; Vousden, K.H. Mutant p53 in cancer: New functions and therapeutic opportunities. *Cancer Cell* **2014**, *25*, 304–317. [CrossRef] [PubMed]
49. Crum, C.P.; McKeon, F.D. *p63* in Epithelial Survival, Germ Cell Surveillance, and Neoplasia. *Annu. Rev. Pathol. Mech. Dis.* **2010**, *5*, 349–371. [CrossRef] [PubMed]
50. Weina, K.; Utikal, J. SOX2 and cancer: Current research and its implications in the clinic. *Clin. Transl. Med.* **2014**, *3*, 19. [CrossRef]
51. Schäfer, M.; Werner, S. Nrf2—A regulator of keratinocyte redox signaling. *Free Radic. Biol. Med.* **2015**, *88*, 243–252. [CrossRef]
52. Kopan, R.; Ilagan, M.X.G. The Canonical Notch Signaling Pathway: Unfolding the Activation Mechanism. *Cell* **2009**, *137*, 216–233. [CrossRef]
53. Sadeqzadeh, E.; De Bock, C.E.; Thorne, R.F. Sleeping Giants: Emerging Roles for the Fat Cadherins in Health and Disease. *Med. Res. Rev.* **2014**, *34*, 190–221. [CrossRef]
54. Doorbar, J. Molecular biology of human papillomavirus infection and cervical cancer. *Clin. Sci.* **2006**, *110*, 525–541. [CrossRef]
55. Biliris, K.A.; Koumantakis, E.; Dokianakis, D.N.; Sourvinos, G.; Spandidos, D.A. Human papillomavirus infection of non-melanoma skin cancers in immunocompetent hosts. *Cancer Lett.* **2000**, *161*, 83–88. [CrossRef]
56. Arbiser, J.L. Implications of Epstein-Barr Virus (EBV)-induced carcinogenesis on cutaneous inflammation and carcinogenesis: Evidence of recurring patterns of angiogenesis and signal transduction. *J. Investig. Dermatol.* **2005**, *124*, xi. [CrossRef] [PubMed]
57. Tolstov, Y.L.; Pastrana, D.V.; Feng, H.; Becker, J.C.; Jenkins, F.J.; Moschos, S.; Chang, Y.; Buck, C.B.; Moore, P.S. Human Merkel cell polyomavirus infection II. MCV is a common human infection that can be detected by conformational capsid epitope immunoassays. *Int. J. Cancer* **2009**, *125*, 1250–1256. [CrossRef] [PubMed]
58. Chen, T.; Hedman, L.; Mattila, P.S.; Jartti, T.; Ruuskanen, O.; Söderlund-Venermo, M.; Hedman, K. Serological evidence of Merkel cell polyomavirus primary infections in childhood. *J. Clin. Virol.* **2011**, *50*, 125–129. [CrossRef] [PubMed]
59. Varga, E.; Kiss, M.; Szabó, K.; Kemény, L. Detection of merkel cell polyomavirus DNA in merkel cell carcinomas. *Br. J. Dermatol.* **2009**, *161*, 930–932. [CrossRef]
60. Sastre-Garau, X.; Peter, M.; Avril, M.F.; Laude, H.; Couturier, J.; Rozenberg, F.; Almeida, A.; Boitier, F.; Carlotti, A.; Couturaud, B.; et al. Merkel cell carcinoma of the skin: Pathological and molecular evidence for a causative role of MCV in oncogenesis. *J. Pathol.* **2009**, *218*, 48–56. [CrossRef]
61. Duncavage, E.J.; Zehnbauer, B.A.; Pfeifer, J.D. Prevalence of Merkel cell polyomavirus in Merkel cell carcinoma. *Mod. Pathol.* **2009**, *22*, 516–521. [CrossRef]
62. Sihto, H.; Kukko, H.; Koljonen, V.; Sankila, R.; Böhling, T.; Joensuu, H. Merkel cell polyomavirus infection, large T antigen, retinoblastoma protein and outcome in Merkel cell carcinoma. *Clin. Cancer Res.* **2011**, *17*, 4806–4813. [CrossRef]
63. Wendzicki, J.A.; Moore, P.S.; Chang, Y. Large T and small T antigens of Merkel cell polyomavirus. *Curr. Opin. Virol.* **2015**, *11*, 38–43. [CrossRef]
64. Verhaegen, M.E.; Mangelberger, D.; Harms, P.W.; Eberl, M.; Wilbert, D.M.; Meireles, J.; Bichakjian, C.K.; Saunders, T.L.; Wong, S.Y.; Dlugosz, A.A. Merkel cell polyomavirus small T antigen initiates merkel cell carcinoma-like tumor development in mice. *Cancer Res.* **2017**, *77*, 3151–3157. [CrossRef]
65. Cheng, J.; Rozenblatt-Rosen, O.; Paulson, K.G.; Nghiem, P.; DeCaprio, J.A. Merkel cell polyomavirus large T antigen has growth-promoting and inhibitory activities. *J. Virol.* **2013**, *87*, 6118–6126. [CrossRef] [PubMed]
66. Shuda, M.; Feng, H.; Kwun, H.J.; Rosen, S.T.; Gjoerup, O.; Moore, P.S.; Chang, Y. T antigen mutations are a human tumor-specific signature for Merkel cell polyomavirus. *Proc. Natl. Acad. Sci. USA* **2008**, *105*, 16272–16277. [CrossRef] [PubMed]
67. Cheng, J.; DeCaprio, J.A.; Fluck, M.M.; Schaffhausen, B.S. Cellular transformation by Simian Virus 40 and Murine Polyoma Virus T antigens. *Semin. Cancer Biol.* **2009**, *19*, 218–228. [CrossRef] [PubMed]

68. Spurgeon, M.E.; Lambert, P.F. Merkel cell polyomavirus: A newly discovered human virus with oncogenic potential. *Virology* **2013**, *435*, 118–130. [CrossRef] [PubMed]
69. Telomere Shortening—An Overview | ScienceDirect Topics. Available online: https://www.sciencedirect.com/topics/neuroscience/telomere-shortening (accessed on 25 March 2020).
70. Richter, T.; von Zglinicki, T. A continuous correlation between oxidative stress and telomere shortening in fibroblasts. *Exp. Gerontol.* **2007**, *42*, 1039–1042. [CrossRef]
71. Nakamura, T.M.; Morin, G.B.; Chapman, K.B.; Weinrich, S.L.; Andrews, W.H.; Lingner, J.; Harley, C.B.; Cech, T.R. Telomerase catalytic subunit homologs from fission yeast and human. *Science* **1997**, *277*, 955–959. [CrossRef]
72. Tsatsakis, A.; Tsoukalas, D.; Fragkiadaki, P.; Vakonaki, E.; Tzatzarakis, M.; Sarandi, E.; Nikitovic, D.; Tsilimidos, G.; Alegakis, A.K. Developing BIOTEL: A Semi-Automated Spreadsheet for Estimating Telomere Length and Biological Age. *Front. Genet.* **2019**, *10*, 84. [CrossRef]
73. Nikolouzakis, T.K.; Vassilopoulou, L.; Fragkiadaki, P.; Sapsakos, T.M.; Papadakis, G.Z.; Spandidos, D.A.; Tsatsakis, A.M.; Tsiaoussis, J. Improving diagnosis, prognosis and prediction by using biomarkers in CRC patients (Review). *Oncol. Rep.* **2018**, *39*, 2455–2472. [CrossRef]
74. Shi, Y.; Zhang, Y.; Zhang, L.; Ma, J.L.; Zhou, T.; Li, Z.X.; Liu, W.D.; Li, W.Q.; Deng, D.J.; You, W.C.; et al. Telomere Length of Circulating Cell-Free DNA and Gastric Cancer in a Chinese Population at High-Risk. *Front. Oncol.* **2019**, *9*, 1434. [CrossRef]
75. Bailey, S.M.; Murnane, J.P. Telomeres, chromosome instability and cancer. *Nucleic Acids Res.* **2006**, *34*, 2408–2417. [CrossRef]
76. Leufke, C.; Leykauf, J.; Krunic, D.; Jauch, A.; Holtgreve-Grez, H.; Böhm-Steuer, B.; Bröcker, E.B.; Mauch, C.; Utikal, J.; Hartschuh, W.; et al. The telomere profile distinguishes two classes of genetically distinct cutaneous squamous cell carcinomas. *Oncogene* **2014**, *33*, 3506–3518. [CrossRef] [PubMed]
77. Caini, S.; Raimondi, S.; Johansson, H.; De Giorgi, V.; Zanna, I.; Palli, D.; Gandini, S. Telomere length and the risk of cutaneous melanoma and non-melanoma skin cancer: A review of the literature and meta-analysis. *J. Dermatol. Sci.* **2015**, *80*, 168–174. [CrossRef] [PubMed]
78. Yamada-Hishida, H.; Nobeyama, Y.; Nakagawa, H. Correlation of telomere length to malignancy potential in non-melanoma skin cancers. *Oncol. Lett.* **2018**, *15*, 393–399. [CrossRef] [PubMed]
79. Wainwright, L.J.; Rees, J.L.; Middleton, P.G. Changes in mean telomere length in basal cell carcinomas of the skin. *Genes Chromosom. Cancer* **1995**, *12*, 45–49. [CrossRef]
80. Han, J.; Qureshi, A.A.; Prescott, J.; Guo, Q.; Ye, L.; Hunter, D.J.; De Vivo, I. A prospective study of telomere length and the risk of skin cancer. *J. Investig. Dermatol.* **2009**, *129*, 415–421. [CrossRef]
81. Anic, G.M.; Sondak, V.K.; Messina, J.L.; Fenske, N.A.; Zager, J.S.; Cherpelis, B.S.; Lee, J.H.; Fulp, W.J.; Epling-Burnette, P.K.; Park, J.Y.; et al. Telomere length and risk of melanoma, squamous cell carcinoma, and basal cell carcinoma. *Cancer Epidemiol.* **2013**, *37*, 434–439. [CrossRef]
82. Liang, G.; Qureshi, A.A.; Guo, Q.; De Vivo, I.; Han, J. No association between telomere length in peripheral blood leukocytes and the risk of nonmelanoma skin cancer. *Cancer Epidemiol. Biomark. Prev.* **2011**, *20*, 1043–1045. [CrossRef]
83. Daniel, M.; Peek, G.W.; Tollefsbol, T.O. Regulation of the human catalytic subunit of telomerase (hTERT). *Gene* **2012**, *498*, 135–146. [CrossRef]
84. Akincilar, S.C.; Unal, B.; Tergaonkar, V. Reactivation of telomerase in cancer. *Cell. Mol. Life Sci.* **2016**, *73*, 1659–1670. [CrossRef]
85. Hoffmeyer, K.; Raggioli, A.; Rudloff, S.; Anton, R.; Hierholzer, A.; Del Valle, I.; Hein, K.; Vogt, R.; Kemler, R. Wnt/β-catenin signaling regulates telomerase in stem cells and cancer cells. *Science* **2012**, *336*, 1549–1554. [CrossRef]
86. Wong, C.W.; Hou, P.S.; Tseng, S.F.; Chien, C.L.; Wu, K.J.; Chen, H.F.; Ho, H.N.; Kyo, S.; Teng, S.C. Krüppel-like transcription factor 4 contributes to maintenance of telomerase activity in stem cells. *Stem Cells* **2010**, *28*, 1510–1517. [CrossRef] [PubMed]
87. Burnworth, B.; Arendt, S.; Muffler, S.; Steinkraus, V.; Bröcker, E.B.; Birek, C.; Hartschuh, W.; Jauch, A.; Boukamp, P. The multi-step process of human skin carcinogenesis: A role for p53, cyclin D1, hTERT, p16, and TSP-1. *Eur. J. Cell Biol.* **2007**, *86*, 763–780. [CrossRef] [PubMed]
88. Ventura, A.; Pellegrini, C.; Cardelli, L.; Rocco, T.; Ciciarelli, V.; Peris, K.; Fargnoli, M.C. Telomeres and telomerase in cutaneous squamous cell carcinoma. *Int. J. Mol. Sci.* **2019**, *20*, 1333. [CrossRef] [PubMed]

89. Parris, C.N.; Jezzard, S.; Silver, A.; MacKie, R.; McGregor, J.M.; Newbold, R.F. Telomerase activity in melanoma and non-melanoma skin cancer. *Br. J. Cancer* **1999**, *79*, 47–53. [CrossRef]
90. Boldrini, L.; Loggini, B.; Gisfredi, S.; Zucconi, Y.; Di Quirico, D.; Biondi, R.; Cervadoro, G.; Barachini, P.; Basolo, F.; Pingitore, R.; et al. Evaluation of telomerase in non-melanoma skin cancer. *Int. J. Mol. Med.* **2003**, *11*, 607–611. [CrossRef]
91. Griewank, K.G.; Murali, R.; Schilling, B.; Schimming, T.; Möller, I.; Moll, I.; Schwamborn, M.; Sucker, A.; Zimmer, L.; Schadendorf, D.; et al. TERT promoter mutations are frequent in cutaneous basal cell carcinoma and squamous cell carcinoma. *PLoS ONE* **2013**, *8*, e80354. [CrossRef]
92. Scott, G.A.; Laughlin, T.S.; Rothberg, P.G. Mutations of the TERT promoter are common in basal cell carcinoma and squamous cell carcinoma. *Mod. Pathol.* **2014**, *27*, 516–523. [CrossRef]
93. Penta, D.; Somashekar, B.S.; Meeran, S.M. Epigenetics of skin cancer: Interventions by selected bioactive phytochemicals. *Photodermatol. Photoimmunol. Photomed.* **2018**, *34*, 42–49. [CrossRef]
94. Smith, Z.D.; Meissner, A. DNA methylation: Roles in mammalian development. *Nat. Rev. Genet.* **2013**, *14*, 204–220. [CrossRef]
95. Rodríguez-Paredes, M.; Bormann, F.; Raddatz, G.; Gutekunst, J.; Lucena-Porcel, C.; Köhler, F.; Wurzer, E.; Schmidt, K.; Gallinat, S.; Wenck, H.; et al. Methylation profiling identifies two subclasses of squamous cell carcinoma related to distinct cells of origin. *Nat. Commun.* **2018**, *9*, 1–9. [CrossRef]
96. Rodríguez-Paredes, M.; Esteller, M. Cancer epigenetics reaches mainstream oncology. *Nat. Med.* **2011**, *17*, 330–339. [CrossRef] [PubMed]
97. Baylin, S.B.; Jones, P.A. A decade of exploring the cancer epigenome-biological and translational implications. *Nat. Rev. Cancer* **2011**, *11*, 726–734. [CrossRef] [PubMed]
98. Falzone, L.; Salemi, R.; Travali, S.; Scalisi, A.; McCubrey, J.A.; Candido, S.; Libra, M. MMP-9 overexpression is associated with intragenic hypermethylation of MMP9 gene in melanoma. *Aging* **2016**, *8*, 933–944. [CrossRef] [PubMed]
99. Candido, S.; Parasiliti Palumbo, G.A.; Pennisi, M.; Russo, G.; Sgroi, G.; Di Salvatore, V.; Libra, M.; Pappalardo, F. EpiMethEx: A tool for large-scale integrated analysis in methylation hotspots linked to genetic regulation. *BMC Bioinform.* **2019**, *19*, 43–53. [CrossRef]
100. Napoli, S.; Scuderi, C.; Gattuso, G.; Di Bella, V.; Candido, S.; Basile, M.S.; Libra, M.; Falzone, L. Functional Roles of Matrix Metalloproteinases and Their Inhibitors in Melanoma. *Cells* **2020**, *9*, 1151. [CrossRef]
101. Hervás-Marín, D.; Higgins, F.; Sanmartín, O.; López-Guerrero, J.A.; Bañó, M.C.; Igual, J.C.; Quilis, I.; Sandoval, J. Genome wide DNA methylation profiling identifies specific epigenetic features in high-risk cutaneous squamous cell carcinoma. *PLoS ONE* **2019**, *14*, e0223341. [CrossRef]
102. Brown, V.L.; Harwood, C.A.; Crook, T.; Cronin, J.G.; Kelsell, D.R.; Proby, C.M. p16INK4a and p14ARF tumor suppressor genes are commonly inactivated in cutaneous squamous cell carcinoma. *J. Investig. Dermatol.* **2004**, *122*, 1284–1292. [CrossRef]
103. Chiles, M.C.; Ai, L.; Zuo, C.; Fan, C.Y.; Smoller, B.R. E-Cadherin Promoter Hypermethylation in Preneoplastic and Neoplastic Skin Lesions. *Mod. Pathol.* **2003**, *16*, 1014–1018. [CrossRef]
104. Murao, K.; Kubo, Y.; Ohtani, N.; Hara, E.; Arase, S. Epigenetic abnormalities in cutaneous squamous cell carcinomas: Frequent inactivation of the RB1/p16 and p53 pathways. *Br. J. Dermatol.* **2006**, *155*, 999–1005. [CrossRef]
105. Takeuchi, T.; Liang, S.B.; Matsuyoshi, N.; Zhou, S.; Miyachi, Y.; Sonobe, H.; Ohtsuki, Y. Loss of T-cadherin (CDH13, H-cadherin) expression in cutaneous squamous cell carcinoma. *Lab. Investig.* **2002**, *82*, 1023–1029. [CrossRef]
106. Venza, I.; Visalli, M.; Tripodo, B.; De Grazia, G.; Loddo, S.; Teti, D.; Venza, M. FOXE1 is a target for aberrant methylation in cutaneous squamous cell carcinoma. *Br. J. Dermatol.* **2010**, *162*, 1093–1097. [CrossRef]
107. Liang, J.; Kang, X.; Halifu, Y.; Zeng, X.; Jin, T.; Zhang, M.; Luo, D.; Ding, Y.; Zhou, Y.; Yakeya, B.; et al. Secreted frizzled-related protein promotors are hypermethylated in cutaneous squamous carcinoma compared with normal epidermis. *BMC Cancer* **2015**, *15*, 641. [CrossRef] [PubMed]
108. Darr, O.A.; Colacino, J.A.; Tang, A.L.; McHugh, J.B.; Bellile, E.L.; Bradford, C.R.; Prince, M.P.; Chepeha, D.B.; Rozek, L.S.; Moyer, J.S. Epigenetic alterations in metastatic cutaneous carcinoma. *Head Neck* **2015**, *37*, 994–1001. [CrossRef] [PubMed]
109. Meier, K.; Drexler, S.K.; Eberle, F.C.; Lefort, K.; Yazdi, A.S. Silencing of ASC in cutaneous squamous cell carcinoma. *PLoS ONE* **2016**, *11*, e0164742. [CrossRef] [PubMed]

110. Nobeyama, Y.; Watanabe, Y.; Nakagawa, H. Silencing of G0/G1 switch gene 2 in cutaneous squamous cell carcinoma. *PLoS ONE* **2017**, *12*, e0187407. [CrossRef] [PubMed]
111. Li, L.; Jiang, M.; Feng, Q.; Kiviat, N.B.; Stern, J.E.; Hawes, S.; Cherne, S.; Lu, H. Aberrant Methylation Changes Detected in Cutaneous Squamous Cell Carcinoma of Immunocompetent Individuals. *Cell Biochem. Biophys.* **2015**, *72*, 599–604. [CrossRef]
112. Toll, A.; Salgado, R.; Espinet, B.; Díaz-Lagares, A.; Hernández-Ruiz, E.; Andrades, E.; Sandoval, J.; Esteller, M.; Pujol, R.M.; Hernández-Muñoz, I. MiR-204 silencing in intraepithelial to invasive cutaneous squamous cell carcinoma progression. *Mol. Cancer* **2016**, *15*, 1. [CrossRef]
113. Venza, M.; Visalli, M.; Catalano, T.; Beninati, C.; Teti, D.; Venza, I. DSS1 promoter hypomethylation and overexpression predict poor prognosis in melanoma and squamous cell carcinoma patients. *Hum. Pathol.* **2017**, *60*, 137–146. [CrossRef]
114. Goldberg, M.; Rummelt, C.; Laerm, A.; Helmbold, P.; Holbach, L.M.; Ballhausen, W.G. Epigenetic silencing contributes to frequent loss of the fragile histidine triad tumour suppressor in basal cell carcinomas. *Br. J. Dermatol.* **2006**, *155*, 1154–1158. [CrossRef]
115. Heitzer, E.; Bambach, I.; Dandachi, N.; Horn, M.; Wolf, P. PTCH promoter methylation at low level in sporadic basal cell carcinoma analysed by three different approaches. *Exp. Dermatol.* **2010**, *19*, 926–928. [CrossRef]
116. Greenberg, E.S.; Chong, K.K.; Huynh, K.T.; Tanaka, R.; Hoon, D.S.B. Epigenetic biomarkers in skin cancer. *Cancer Lett.* **2014**, *342*, 170–177. [CrossRef] [PubMed]
117. Harms, P.W.; Harms, K.L.; Moore, P.S.; DeCaprio, J.A.; Nghiem, P.; Wong, M.K.K.; Brownell, I. The biology and treatment of Merkel cell carcinoma: Current understanding and research priorities. *Nat. Rev. Clin. Oncol.* **2018**, *15*, 763–776. [CrossRef] [PubMed]
118. Salemi, R.; Falzone, L.; Madonna, G.; Polesel, J.; Cinà, D.; Mallardo, D.; Ascierto, P.A.; Libra, M.; Candido, S. MMP-9 as a Candidate Marker of Response to BRAF Inhibitors in Melanoma Patients With BRAFV600E Mutation Detected in Circulating-Free DNA. *Front. Pharmacol.* **2018**, *9*, 856. [CrossRef] [PubMed]
119. Kouzarides, T. Chromatin Modifications and Their Function. *Cell* **2007**, *128*, 693–705. [CrossRef]
120. Liang, G.; Lin, J.C.Y.; Wei, V.; Yoo, C.; Cheng, J.C.; Nguyen, C.T.; Weisenberger, D.J.; Egger, G.; Takai, D.; Gonzales, F.A.; et al. Distinct localization of histone H3 acetylation and H3-K4 methylation to the transcription start sites in the human genome. *Proc. Natl. Acad. Sci. USA* **2004**, *101*, 7357–7362. [CrossRef]
121. Nandakumar, V.; Vaid, M.; Katiyar, S.K. (−)-Epigallocatechin-3-gallate reactivates silenced tumor suppressor genes, Cip1/p21 and p 16 INK4a, by reducing DNA methylation and increasing histones acetylation in human skin cancer cells. *Carcinogenesis* **2011**, *32*, 537–544. [CrossRef]
122. Smits, M.; Van Rijn, S.; Hulleman, E.; Biesmans, D.; Van Vuurden, D.G.; Kool, M.; Haberler, C.; Aronica, E.; Vandertop, W.P.; Noske, D.P.; et al. EZH2-regulated DAB2IP is a medulloblastoma tumor suppressor and a positive marker for survival. *Clin. Cancer Res.* **2012**, *18*, 4048–4058. [CrossRef]
123. Rao, R.C.; Chan, M.P.; Andrews, C.A.; Kahana, A. EZH2, proliferation rate, and aggressive tumor subtypes in cutaneous basal cell carcinoma. *JAMA Oncol.* **2016**, *2*, 962–963. [CrossRef]
124. Rao, R.C.; Chan, M.P.; Andrews, C.A.; Kahana, A. Epigenetic markers in basal cell carcinoma: Universal themes in oncogenesis and tumor stratification?—A short report. *Cell. Oncol.* **2018**, *41*, 693–698. [CrossRef]
125. Harms, K.L.; Chubb, H.; Zhao, L.; Fullen, D.R.; Bichakjian, C.K.; Johnson, T.M.; Carskadon, S.; Palanisamy, N.; Harms, P.W. Increased expression of EZH2 in Merkel cell carcinoma is associated with disease progression and poorer prognosis. *Hum. Pathol.* **2017**, *67*, 78–84. [CrossRef]
126. Orouji, E.; Utikal, J. Tackling malignant melanoma epigenetically: Histone lysine methylation. *Clin. Epigenetics* **2018**, *10*, 145. [CrossRef] [PubMed]
127. Raman, A.T.; Rai, K. Loss of histone acetylation and H3K4 methylation promotes melanocytic malignant transformation. *Mol. Cell. Oncol.* **2018**, *5*, e1359229. [CrossRef] [PubMed]
128. Wightman, B.; Ha, I.; Ruvkun, G. Posttranscriptional regulation of the heterochronic gene lin-14 by lin-4 mediates temporal pattern formation in C. elegans. *Cell* **1993**, *75*, 855–862. [CrossRef]
129. Lee, R.C.; Feinbaum, R.L.; Ambros, V. The C. elegans heterochronic gene lin-4 encodes small RNAs with antisense complementarity to lin-14. *Cell* **1993**, *75*, 843–854. [CrossRef]
130. Ha, M.; Kim, V.N. Regulation of microRNA biogenesis. *Nat. Rev. Mol. Cell Biol.* **2014**, *15*, 509–524. [CrossRef] [PubMed]
131. Broughton, J.P.; Lovci, M.T.; Huang, J.L.; Yeo, G.W.; Pasquinelli, A.E. Pairing beyond the Seed Supports MicroRNA Targeting Specificity. *Mol. Cell* **2016**, *64*, 320–333. [CrossRef]

132. Vasudevan, S. Posttranscriptional Upregulation by MicroRNAs. *Wiley Interdiscip. Rev. RNA* **2012**, *3*, 311–330. [CrossRef]
133. Makarova, J.A.; Shkurnikov, M.U.; Wicklein, D.; Lange, T.; Samatov, T.R.; Turchinovich, A.A.; Tonevitsky, A.G. Intracellular and extracellular microRNA: An update on localization and biological role. *Prog. Histochem. Cytochem.* **2016**, *51*, 33–49. [CrossRef]
134. O'Brien, J.; Hayder, H.; Zayed, Y.; Peng, C. Overview of microRNA biogenesis, mechanisms of actions, and circulation. *Front. Endocrinol.* **2018**, *9*, 402. [CrossRef]
135. Creemers, E.E.; Tijsen, A.J.; Pinto, Y.M. Circulating MicroRNAs: Novel biomarkers and extracellular communicators in cardiovascular disease? *Circ. Res.* **2012**, *110*, 483–495. [CrossRef]
136. Mitchell, P.S.; Parkin, R.K.; Kroh, E.M.; Fritz, B.R.; Wyman, S.K.; Pogosova-Agadjanyan, E.L.; Peterson, A.; Noteboom, J.; O'Briant, K.C.; Allen, A.; et al. Circulating microRNAs as stable blood-based markers for cancer detection. *Proc. Natl. Acad. Sci. USA* **2008**, *105*, 10513–10518. [CrossRef] [PubMed]
137. Doukas, S.G.; Vageli, D.P.; Lazopoulos, G.; Spandidos, D.A.; Sasaki, C.T.; Tsatsakis, A. The Effect of NNK, A Tobacco Smoke Carcinogen, on the miRNA and Mismatch DNA Repair Expression Profiles in Lung and Head and Neck Squamous Cancer Cells. *Cells* **2020**, *9*, 1031. [CrossRef] [PubMed]
138. Silantyev, A.; Falzone, L.; Libra, M.; Gurina, O.; Kardashova, K.; Nikolouzakis, T.; Nosyrev, A.; Sutton, C.; Mitsias, P.; Tsatsakis, A. Current and Future Trends on Diagnosis and Prognosis of Glioblastoma: From Molecular Biology to Proteomics. *Cells* **2019**, *8*, 863. [CrossRef] [PubMed]
139. Filetti, V.; Falzone, L.; Rapisarda, V.; Caltabiano, R.; Eleonora Graziano, A.C.; Ledda, C.; Loreto, C. Modulation of microRNA expression levels after naturally occurring asbestiform fibers exposure as a diagnostic biomarker of mesothelial neoplastic transformation. *Ecotoxicol. Environ. Saf.* **2020**, *198*, 110640. [CrossRef] [PubMed]
140. Falzone, L.; Lupo, G.; La Rosa, G.R.M.; Crimi, S.; Anfuso, C.D.; Salemi, R.; Rapisarda, E.; Libra, M.; Candido, S. Identification of Novel MicroRNAs and Their Diagnostic and Prognostic Significance in Oral Cancer. *Cancers* **2019**, *11*, 610. [CrossRef]
141. Falzone, L.; Romano, G.L.; Salemi, R.; Bucolo, C.; Tomasello, B.; Lupo, G.; Anfuso, C.D.; Spandidos, D.A.; Libra, M.; Candido, S. Prognostic significance of deregulated microRNAs in uveal melanomas. *Mol. Med. Rep.* **2019**, *19*, 2599–2610. [CrossRef]
142. Falzone, L.; Scola, L.; Zanghì, A.; Biondi, A.; Di Cataldo, A.; Libra, M.; Candido, S. Integrated analysis of colorectal cancer microRNA datasets: Identification of microRNAs associated with tumor development. *Aging* **2018**, *10*, 1000–1014. [CrossRef]
143. Falzone, L.; Candido, S.; Salemi, R.; Basile, M.S.; Scalisi, A.; McCubrey, J.A.; Torino, F.; Signorelli, S.S.; Montella, M.; Libra, M. Computational identification of microRNAs associated to both epithelial to mesenchymal transition and NGAL/MMP-9 pathways in bladder cancer. *Oncotarget* **2016**, *7*, 72758–72766. [CrossRef]
144. Hafsi, S.; Candido, S.; Maestro, R.; Falzone, L.; Soua, Z.; Bonavida, B.; Spandidos, D.A.; Libra, M. Correlation between the overexpression of Yin Yang 1 and the expression levels of miRNAs in Burkitt's lymphoma: A computational study. *Oncol. Lett.* **2016**, *11*, 1021–1025. [CrossRef]
145. Sand, M.; Bechara, F.G.; Gambichler, T.; Sand, D.; Friedländer, M.R.; Bromba, M.; Schnabel, R.; Hessam, S. Next-generation sequencing of the basal cell carcinoma miRNome and a description of novel microRNA candidates under neoadjuvant vismodegib therapy: An integrative molecular and surgical case study. *Ann. Oncol. Off. J. Eur. Soc. Med. Oncol.* **2016**, *27*, 332–338. [CrossRef]
146. Yi, R.; Poy, M.N.; Stoffel, M.; Fuchs, E. A skin microRNA promotes differentiation by repressing "stemness". *Nature* **2008**, *452*, 225–229. [CrossRef] [PubMed]
147. Schnidar, H.; Eberl, M.; Klingler, S.; Mangelberger, D.; Kasper, M.; Hauser-Kronberger, C.; Regl, G.; Kroismayr, R.; Moriggl, R.; Sibilia, M.; et al. Epidermal growth factor receptor signaling synergizes with hedgehog/GLI in oncogenic transformation via activation of the MEK/ERK/JUN pathway. *Cancer Res.* **2009**, *69*, 1284–1292. [CrossRef] [PubMed]
148. Sonkoly, E.; Lovén, J.; Xu, N.; Meisgen, F.; Wei, T.; Brodin, P.; Jaks, V.; Kasper, M.; Shimokawa, T.; Harada, M.; et al. MicroRNA-203 functions as a tumor suppressor in basal cell carcinoma. *Oncogenesis* **2012**, *1*, e3. [CrossRef] [PubMed]
149. Hu, P.; Ma, L.; Wu, Z.; Zheng, G.; Li, J. Expression of miR-34a in basal cell carcinoma patients and its relationship with prognosis. *J. BUON.* **2019**, *24*, 1283–1288. [PubMed]

150. Mizrahi, A.; Barzilai, A.; Gur-Wahnon, D.; Ben-Dov, I.Z.; Glassberg, S.; Meningher, T.; Elharar, E.; Masalha, M.; Jacob-Hirsch, J.; Tabibian-Keissar, H.; et al. Alterations of microRNAs throughout the malignant evolution of cutaneous squamous cell carcinoma: The role of miR-497 in epithelial to mesenchymal transition of keratinocytes. *Oncogene* **2018**, *37*, 218–230. [CrossRef]
151. Yu, X.; Li, Z. The role of miRNAs in cutaneous squamous cell carcinoma. *J. Cell. Mol. Med.* **2016**, *20*, 3–9. [CrossRef]
152. García-Sancha, N.; Corchado-Cobos, R.; Pérez-Losada, J.; Cañueto, J. MicroRNA dysregulation in cutaneous squamous cell carcinoma. *Int. J. Mol. Sci.* **2019**, *20*, 2181.
153. Cañueto, J.; Cardeñoso-Álvarez, E.; García-Hernández, J.L.; Galindo-Villardón, P.; Vicente-Galindo, P.; Vicente-Villardón, J.L.; Alonso-López, D.; De Las Rivas, J.; Valero, J.; Moyano-Sanz, E.; et al. MicroRNA (miR)-203 and miR-205 expression patterns identify subgroups of prognosis in cutaneous squamous cell carcinoma. *Br. J. Dermatol.* **2017**, *177*, 168–178. [CrossRef]
154. Stojadinovic, O.; Ramirez, H.; Pastar, I.; Gordon, K.A.; Stone, R.; Choudhary, S.; Badiavas, E.; Nouri, K.; Tomic-Canic, M. MiR-21 and miR-205 are induced in invasive cutaneous squamous cell carcinomas. *Arch. Dermatol. Res.* **2017**, *309*, 133–139. [CrossRef]
155. Gong, Z.H.; Zhou, F.; Shi, C.; Xiang, T.; Zhou, C.K.; Wang, Q.Q.; Jiang, Y.S.; Gao, S.F. miRNA-221 promotes cutaneous squamous cell carcinoma progression by targeting PTEN. *Cell. Mol. Biol. Lett.* **2019**, *24*, 9. [CrossRef]
156. Zhang, L.; Xiang, P.; Han, X.; Wu, L.; Li, X.; Xiong, Z. Decreased expression of microRNA-20a promotes tumor progression and predicts poor prognosis of cutaneous squamous cell carcinoma. *Int. J. Clin. Exp. Pathol.* **2015**, *8*, 11446–11451. [PubMed]
157. Ning, M.S.; Kim, A.S.; Prasad, N.; Levy, S.E.; Zhang, H.; Andl, T. Characterization of the Merkel Cell Carcinoma miRNome. *J. Skin Cancer* **2014**, *2014*, 289548. [CrossRef] [PubMed]
158. Veija, T.; Sahi, H.; Koljonen, V.; Bohling, T.; Knuutila, S.; Mosakhani, N. miRNA-34a underexpressed in Merkel cell polyomavirus-negative Merkel cell carcinoma. *Virchows Arch.* **2015**, *466*, 289–295. [CrossRef] [PubMed]
159. Renwick, N.; Cekan, P.; Masry, P.A.; McGeary, S.E.; Miller, J.B.; Hafner, M.; Li, Z.; Mihailovic, A.; Morozov, P.; Brown, M.; et al. Multicolor microRNA FISH effectively differentiates tumor types. *J. Clin. Investig.* **2013**, *123*, 2694–2702. [CrossRef]
160. Konstatinell, A.; Coucheron, D.H.; Sveinbjørnsson, B.; Moens, U. MicroRNAs as potential biomarkers in merkel cell carcinoma. *Int. J. Mol. Sci.* **2018**, *19*, 1873.
161. Tuaeva, N.O.; Falzone, L.; Porozov, Y.B.; Nosyrev, A.E.; Trukhan, V.M.; Kovatsi, L.; Spandidos, D.A.; Drakoulis, N.; Kalogeraki, A.; Mamoulakis, C.; et al. Translational Application of Circulating DNA in Oncology: Review of the Last Decades Achievements. *Cells* **2019**, *8*, 1251. [CrossRef]
162. Xie, H.; Lee, L.; Caramuta, S.; Höög, A.; Browaldh, N.; Björnhagen, V.; Larsson, C.; Lui, W.O. MicroRNA expression patterns related to merkel cell polyomavirus infection in human Merkel cell carcinoma. *J. Investig. Dermatol.* **2014**, *134*, 507–517. [CrossRef]
163. Lu, T.; Chen, S.; Qu, L.; Wang, Y.; Chen, H.D.; He, C. Identification of a five-miRNA signature predicting survival in cutaneous melanoma cancer patients. *PeerJ* **2019**, *2019*, e7831. [CrossRef]
164. Hanniford, D.; Zhong, J.; Koetz, L.; Gaziel-Sovran, A.; Lackaye, D.J.; Shang, S.; Pavlick, A.; Shapiro, R.; Berman, R.; Darvishian, F.; et al. A miRNA-based signature detected in primary melanoma tissue predicts development of brain metastasis. *Clin. Cancer Res.* **2015**, *21*, 4903–4912. [CrossRef]
165. Fenech, M.; Kirsch-Volders, M.; Natarajan, A.T.; Surralles, J.; Crott, J.W.; Parry, J.; Norppa, H.; Eastmond, D.A.; Tucker, J.D.; Thomas, P. Molecular mechanisms of micronucleus, nucleoplasmic bridge and nuclear bud formation in mammalian and human cells. *Mutagenesis* **2011**, *26*, 125–132. [CrossRef]
166. Mateuca, R.; Lombaert, N.; Aka, P.V.; Decordier, I.; Kirsch-Volders, M. Chromosomal changes: Induction, detection methods and applicability in human biomonitoring. *Biochimie* **2006**, *88*, 1515–1531. [CrossRef] [PubMed]
167. Pardini, B.; Viberti, C.; Naccarati, A.; Allione, A.; Oderda, M.; Critelli, R.; Preto, M.; Zijno, A.; Cucchiarale, G.; Gontero, P.; et al. Increased micronucleus frequency in peripheral blood lymphocytes predicts the risk of bladder cancer. *Br. J. Cancer* **2017**, *116*, 202–210. [CrossRef] [PubMed]
168. Nikolouzakis, T.K.; Stivaktakis, P.D.; Apalaki, P.; Kalliantasi, K.; Sapsakos, T.M.; Spandidos, D.A.; Tsatsakis, A.; Souglakos, J.; Tsiaoussis, J. Effect of systemic treatment on the micronuclei frequency in the peripheral blood of patients with metastatic colorectal cancer. *Oncol. Lett.* **2019**, *17*, 2703–2712. [CrossRef] [PubMed]

169. Diem, C.; Rünger, T.M. Processing of three different types of DNA damage in cell lines of a cutaneous squamous cell carcinoma progression model. *Carcinogenesis* **1997**, *18*, 657–662. [CrossRef] [PubMed]
170. Emri, G.; Wenczl, E.; Van Erp, P.; Jans, J.; Roza, L.; Horkay, I.; Schothorst, A.A. Low doses of UVB or UVA induce chromosomal aberrations in cultured human skin cells. *J. Investig. Dermatol.* **2000**, *115*, 435–440. [CrossRef] [PubMed]
171. Sanford, K.K.; Parshad, R.; Price, F.M.; Tarone, R.E.; Thompson, J.; Guerry, D. Radiation-induced chromatid breaks and DNA repair in blood lymphocytes of patients with dysplastic nevi and/or cutaneous melanoma. *J. Investig. Dermatol.* **1997**, *109*, 546–549. [CrossRef]

© 2020 by the authors. Licensee MDPI, Basel, Switzerland. This article is an open access article distributed under the terms and conditions of the Creative Commons Attribution (CC BY) license (http://creativecommons.org/licenses/by/4.0/).

Review

Recent Advances in Signaling Pathways Comprehension as Carcinogenesis Triggers in Basal Cell Carcinoma

Mircea Tampa [1,2], Simona Roxana Georgescu [1,2,*], Cristina Iulia Mitran [3], Madalina Irina Mitran [3], Clara Matei [1], Cristian Scheau [4], Carolina Constantin [5,6] and Monica Neagu [5,6,7]

1. Department of Dermatology, "Carol Davila" University of Medicine and Pharmacy, 050474 Bucharest, Romania; tampa_mircea@yahoo.com (M.T.); matei_clara@yahoo.com (C.M.)
2. Department of Dermatology, "Victor Babes" Clinical Hospital for Infectious Diseases, 030303 Bucharest, Romania
3. Department of Microbiology, "Carol Davila" University of Medicine and Pharmacy, 050474 Bucharest, Romania; cristina.mitran@drd.umfcd.ro (C.I.M.); irina.mitran@drd.umfcd.ro (M.I.M.)
4. Department of Physiology, "Carol Davila" University of Medicine and Pharmacy, 050474 Bucharest, Romania; cristian.scheau@umfcd.ro
5. Immunology Department, "Victor Babes" National Institute of Pathology, 050096 Bucharest, Romania; caroconstantin@gmail.com (C.C.); neagu.monica@gmail.com (M.N.)
6. Colentina Clinical Hospital, 020125 Bucharest, Romania
7. Faculty of Biology, University of Bucharest, 76201 Bucharest, Romania
* Correspondence: srg.dermatology@gmail.com

Received: 18 July 2020; Accepted: 16 September 2020; Published: 18 September 2020

Abstract: Basal cell carcinoma (BCC) is the most common malignant skin tumor. BCC displays a different behavior compared with other neoplasms, has a slow evolution, and metastasizes very rarely, but sometimes it causes an important local destruction. Chronic ultraviolet exposure along with genetic factors are the most important risk factors involved in BCC development. Mutations in the *PTCH1* gene are associated with Gorlin syndrome, an autosomal dominant disorder characterized by the occurrence of multiple BCCs, but are also the most frequent mutations observed in sporadic BCCs. PTCH1 encodes for PTCH1 protein, the most important negative regulator of the Hedgehog (Hh) pathway. There are numerous studies confirming Hh pathway involvement in BCC pathogenesis. Although Hh pathway has been intensively investigated, it remains incompletely elucidated. Recent studies on BCC tumorigenesis have shown that in addition to Hh pathway, there are other signaling pathways involved in BCC development. In this review, we present recent advances in BCC carcinogenesis.

Keywords: basal cell carcinoma; Hedgehog pathway; signaling pathways; carcinogenesis

1. Introduction

Basal cell carcinoma (BCC) is the most prevalent form of skin cancer, developing on sun exposed areas, especially in the fourth decade of life. BCC is a slow-growing, locally invasive tumor, with a low capacity of metastatic spread [1,2]. It is commonly recognized that only 0.0028–0.55% of BCCs will metastasize [3]. Exposure to ultraviolet (UV) light is a key factor in its pathogenesis [4]. Therefore, most cases of BCC are diagnosed in individuals with fair skin phototypes that carry out activities involving intense, intermittent or continuous exposure to UV [5]. In addition, the exposure to ionizing radiation, arsenic or coal tar derivatives increases the risk of developing a BCC. The incidence of BCC is higher compared to the general population in two particular scenarios, immunosuppressed patients and patients with certain genodermatoses such as Gorlin syndrome [6,7]. In this review, we focused on recent advances related to the signaling pathways involved in BCC carcinogenesis.

2. The Genetic Basis of Basal Cell Carcinoma Initiation and Therapy Resistance

2.1. Genes Involved in Nevoid Basal Cell Carcinoma Syndrome

Nevoid basal cell carcinoma syndrome (NBCCS) is an autosomal dominant disorder characterized by mutations in the *patched (PTCH)1* gene, *PTCH2* gene and *suppressor of the fused (SUFU)* gene, which are negative regulators of the hedgehog (Hh) pathway [8]. *PTCH1*, located on chromosome 9q22.3, encodes the homologous transmembrane protein PTCH1 that acts as a receptor for the Hh pathway [9]. *PTCH2* is located on chromosome 1p34 and encodes for PTCH2 and *SUFU* is located on chromosome 10q24.32 and encodes for the suppressor of a fused homologous protein, SUFU [8]. The prevalence of NBCCS ranges between 1/57.000 and 1/256.000. It is a multisystemic disease characterized by the development of multiple BCCs, jaw keratocysts, palmar and plantar pits, abnormalities of the bones and eyes, cardiac dysfunction, calcification of the falx cerebri, etc. In about 5% of cases patients can associate intellectual deficiency [10]. NBCCS is also known as Gorlin Goltz syndrome, after the name of those who described it as a distinct entity in 1960. It occurs most commonly in Caucasian adults aged 17–35 years, with no sex predominance [11].

2.2. Genes Involved in Sporadic BCC

Sporadic BCC is also related to genetic alterations in components of the Hh pathway. Mutations in the *PTCH1* gene were observed in 30–60% of cases, in the *smoothened (SMO)* gene in 10–20%, and to a lesser extent in the *SUFU* gene. Alterations involving glioma-associated oncogenes (*GLI*) are rare [12]. Mutations in *TP53* gene were observed in a high number of cases, over 50% of BCCs. *TP53* encodes for the p53 protein, one of the most important regulators of the cell cycle. Mutations in *TP53* seem to be involved in the initiation of the malignant process but also in tumor progression [13].

Given that BCC has a great diversity in terms of clinical appearance, histopathological forms, evolution and response to treatment, Bonilla et al. considered that there are many other genes involved in its pathogenesis. Thus, they identified mutations in *MYCN, PTPN14,* and *LATS1. MYCN* alterations were observed in 30% of the studied BCC samples, most of them being identified in the Myc box 1 (MB1) region. Mutations in *PTPN14* were observed in 23% of cases, and in *LATS1* in 16% of cases [14].

Moreover, alterations in pigmentary genes were detected in BCC patients [13]. Genetic studies have revealed several BCC susceptibility regions such as 1p36, 1q42, 5p13.3, 5p15, and 12q11-13. A recent study has found new susceptibility regions on chromosome 5, 5q11.2-14.3, 5q22.1-23.3, and 5q31-35.3. These findings may underlie the development of new diagnostic tools and therapeutic approaches in BCC management [15].

A large number of mutations have been revealed so far in BCC cells, therefore Jayaraman et al. hypothesized that this variety of mutations leads to the activation of the host's defense system, which may explain why BCC evolves slowly and metastasizes very rarely [16]. In line with this, the study by Dai et al. performed on 19 BCC samples has revealed the overexpression of 222 genes and the downregulation of 91 genes. Upregulated genes were involved in cell cycle regulation and mitosis, while downregulated genes were involved in cell differentiation and unsaturated fatty acid metabolism. The increased expression of cyclin-dependent kinase (CDK)-1, a regulator of the cell cycle, has been observed, and may represent a novel target for new therapies in BCC [17].

2.3. Genes Linked to Therapy Resistance

There are several attempts to target Hh pathway, some of them already approved in BCC, some of them in the preclinical phase (Table 1). Gene mutations plays an important role in the response to drug therapy. About 20% of patients with BCC treated with vismodegib, a SMO inhibitor, undergo treatment failure within one year of treatment. The main mechanism involved in resistance development is the overexpression of several components of the Hh pathway [18]. Mutations were observed in both the vicinity and distally of drug binding situs of SMO. SMO mutations that occur in the vicinity of the drug binding domain such as D473, H231, W281, Q477, V321, I408, and C469, have been detected only in resistant BCCs which suggests the role of the drug therapy in acquiring these mutations. Mutations distal to the drug binding domain such as T241M, A459V, L412F, S533N, and W535L were found in both untreated BCCs and resistant tumors revealing their inherent role [19].

Table 1. Hedgehog inhibitors in basal cell carcinoma.

Target	Therapy Molecule	Reference
SMO	Vismodegib *	Sekulic et al. [20]
SMO	Sonidegib **	Danial et al. [21]
SMO	Itraconazole	Kim et al. [22]
SMO	BMS-833923	Siu et al. [23]
SMO	Taladegib	Bendell et al. [24]
SMO	Patidegib	Jimeno et al. [25]
SMO	NVP-LQ506	Peukert et al. [26]
SHH	Robotnikinin	Hassounah et al. [27]
GLI	GANT-58 and GANT-61	Lauth et al. [28]
GLI	Arsenic trioxide	Ally et al. [29]

* Approved by the FDA in 2012, ** Approved by the FDA in 2015.

Vismodegib resistance in BCC was also linked to mutations in *TP53* [18]. Mutations in *SUFU* were linked to resistance to vismodegib in a small number of cases [3]. However, tumor resistance was identified in patients without an identifiable mutation [30]. Secondary resistance to vismodegib was first described in a patient diagnosed with medulloblastoma [31].

The mechanisms involved in BCC resistance are not only related to mutations in the canonical Hh pathway, therefore Whitson et al. have shown in a mouse model that the activation of non-canonical Hh pathway by MKL1/SRF is related to the resistance to SMO inhibitors in some BCCs. Thus, they have highlighted the role of MKL1 inhibitors in the treatment of BCC in combination with SMO inhibitors, MKL1 inhibitors could exhibit a synergistic effect [32].

3. Hedgehog Pathway—From Discovery to New Concepts

The Hh pathway plays an essential role in human embryogenesis, being involved in cell differentiation, cell growth, and morphogenesis. [33]. Under normal conditions, the hair follicle and the skin are the only two regions where the Hh signaling displays post-embryonic activity. Hh signaling is also active in stem cells and in tissues undergoing regeneration, having an important role in wound healing [13]. The ectopic activation of the Hh pathway contributes to tumorigenesis, metastasis and resistance to therapy [34]. The first link between BCC and the Hh pathway was revealed in the context of the discovery of loss-of-function mutations in *PTCH1* gene in patients with Gorlin syndrome [35].

Recent research has revealed that Hh signaling can be activated through different pathways [36]. Thus, Hh signaling was classified as canonical and non-canonical [37]. The canonical Hh pathway involves a GLI-mediated transcription. When the activation of Hh pathway occurs independently of GLI-mediated transcription it is categorized as non-canonical Hh pathway [38]. The aberrant stimulation of the Hh pathway as a result of mutations in *PTCH1* and *SMO* is involved in the development of BCC [39]. The binding of one of the Hh ligands to PTCH1, a 12-pass transmembrane receptor protein that prevents the activation of Hh pathway, is the first step required for the activation of the canonical Hh pathway. In vertebrates, three ligands were described, including Sonic hedgehog (Shh), Indian hedgehog (Ihh), and Desert hedgehog (Dhh), of which Shh is the strongest pathway activator [37]. Hh bindings proteins, such as Hh interacting proteins, sequester Hh ligands and in this manner control the amount of Shh that binds to PTCH1 [40]. The Hh ligands bind to PTCH1 and remove it from the primary cilium resulting in the stimulation of SMO, a 7-pass transmembrane protein, and its translocation to the primary cilium. The accumulation of SMO triggers a cascade of events that promote the transcriptional activation of GLI, resulting in cell proliferation (Figure 1) [39,41].

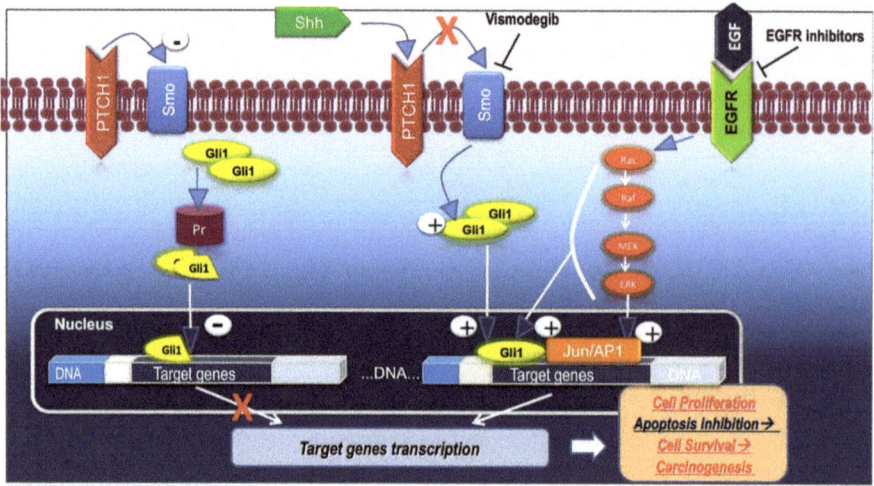

Figure 1. Hh pathway (inactive state and active state) and the crosstalk between Hh and EGFR pathways. Shh—Sonic hedgehog; PTCH1—protein patched homolog 1; SMO—smoothened protein; GLI—glioma-associated oncogenes; Pr—proteasome; EGFR—epidermal growth factor receptor; Ras—rat sarcoma virus; Raf—rapidly accelerated fibrosarcoma; MEK—mitogen-activated protein kinase kinase; ERK—extracellular signal-regulated kinase.

Recent studies have shown that there are two different categories of non-canonical Hh signaling, type 1 acting via PTCH1, in a SMO independent manner and type 2 acting via SMO, independently of GLI regulation. The role of non-canonical Hh signaling in skin cancers is not fully elucidated [38].

Non-canonical pathways involved in BCC tumorigenesis include K-Ras, transforming growth factor-β (TGF-β), PI3K/Protein kinase B (AKT)/mammalian target of rapamycin (mTOR), protein kinase C, and the serum-response factor-megakaryoblastic leukemia-1 pathway [34,37].

In the absence of Hh ligands, PTCH1 is located on the primary cilium and does not allow SMO migration and insertion into the primary cilium. The GLI transcription factors are phosphorylated by protein kinases and undergo proteolytic cleavage resulting in repressor molecules that will suppress the activation of the Hh pathway [42,43]. In other words, Hh ligands are the initiators of Hh pathway, PTCH1 operates as a negative regulatory receptor, and SMO as a positive regulatory receptor. In the absence of Hh ligands, PTCH1 binds to SMO preventing the pathway activation [13,44]. GLI transcription factors are blocked into the cytoplasm by various proteins acting as mediators. The most important proteins involved are protein kinase A (PKA) and SUFU. The proteolytic cleavage of GLI transcription factors generates the repressor forms, GLI2R and GLI3R. The mechanism by which PTCH1 suppresses SMO function in the absence of ligands is not fully known. There have been postulated several theories. It seems that PTCH1 does not allow SMO activation by blocking SMO agonists, such as primary cilium oxysterols. Another theory claims that PTCH1 increases the influx of SMO antagonists into the primary cilium [45].

4. The Role of Inflammation and Immune Response in BCC Pathogenesis

A substantial body of evidence indicates that BCC is an immunogenic tumor. This is supported by the increased incidence of BCC among immunosuppressed subjects and by the presence of numerous immune cells that infiltrate the tumor and peritumoral area [46].

4.1. Tumor Microenvironment in BCC

In BCC samples, it has been observed a high number of immature CD11c + myeloid dendritic cells (DCs) which suggests that tumor microenvironment has an immunosuppressive effect, knowing that immature DCs induce downregulation of T cells [47]. Regulatory T cells (Tregs) have the ability to prevent the maturation of DCs [48] and it should be highlighted that an increased number of FOXP3 + Treg have been identified in the tumor and peritumoral area, and not identified in normal skin. However, the role of Tregs in BCC is not fully understood. [49]. The cell composition of the inflammatory infiltrate may be influenced by various factors, UV exposure or treatment. It has been shown that, after exposure to UV light, Langerhans cells stimulate the function of the Th2 subset of CD4+ T cells [50]. After immunocryosurgery, a significant increase in the CD3+/Foxp3+ ratio was observed, which denotes the induction of an antitumor response [51]. Tumor-associated macrophages (TAMs) also exert an immunosuppressive effect in BCC, their presence being associated with increased invasion capacity and high microvessel density [47].

The increased expression of Th2 cytokines is an additional factor in the generation of an immunosuppressive tumor microenvironment. High levels of IL-4 and IL-10, type 2 cytokines, have been identified in BCC [48]. High levels of IL-10 correlate with a decreased expression of MHC-I and other molecules such as ICAM-1, CD40, CD80, and HLA-ABC. In BCC the low number of CD8 cells and decreased expression of MHC-I allow the tumor escape from immune surveillance. Treatment with Hh inhibitors is associated with an increase in the number of CD8+ and CD4+ T cells [52]. A recent study suggests a potential role of IL-23/Th17-related cytokines in BCC. The role of IL-23 in carcinogenesis is not fully elucidated, increased levels of IL-23 being associated with both tumor growth and apoptosis. In contrast, in regressing BCC, a Th1 immune response has been revealed [53].

4.2. The Link between Inflammation and Hh Signaling

Recent research highlights the link between Hh signaling and immune cells. Data obtained from a study conducted on murine BCC cells, have revealed that Hh signaling induces the migration of myeloid-derived suppressor cells (MDSC) and M2 polarization of macrophages, resulting in an immunosuppressive tumor microenvironment. Moreover, keratinocytes presenting SMO oncogene release TGF-β with an inhibitory action on effector T cells. Another immune-related mechanism in which Hh signaling is involved is the decrease in MHC-I molecule expression on the cell membrane of malignant cells, a phenomenon that hinders immune system recognition [54]. A recent study has shown that the inhibition of the Hh pathway in BCC patients treated with vismodegib or sonidegib (SMO inhibitors) resulted in an increased MHC-I expression in tumor cells associated with a high number of CD4 and CD8 T cells infiltrating the tumors [55]. Given the immunosuppressive effect of Hh signaling, it was found useful to associate Hh inhibitors with immune checkpoint inhibitors. Thus, patients with BCC who received nivolumab or pembrolizumab obtained encouraging results [54]. Hh signaling seems to reduce TCR signaling in mature T cells, and the inhibition of Hh signaling promotes T cell activation and proliferation and hence induces an anti-tumoral effect [55]. The relationship between inflammation and carcinogenesis has been intensively studied in the last decade [50,56]. Chronic inflammation is a key factor in cell malignant transformation [57]. Pro-inflammatory cytokines are important players in the initiation and perpetuation of the inflammatory process [58]. Interleukin 6 (IL-6) is the pro-inflammatory cytokine prototype [59,60]. A recent study has shown that IL-6 stimulates tumorigenesis by synergistically acting with Hh pathway. Hh—IL-6 signaling tandem is based on the activation of the signal transducer and activator of transcription 3 (STAT3) via IL-6/Jak2 pathway. IL-6 and Hh pathway interact at the level of cis-regulatory regions following the cooperation of GLI and STAT3. Regarding the activation of IL-6 signaling, three mechanisms have been suggested, including its activation by Hh pathway, its activation under the influence of the tumor microenvironment or via sIL6R-mediated trans-signaling [61].

5. Crosstalk between Hh Signaling Pathway and Other Signaling Pathways in BCC

BCC carcinogenesis is orchestrated by various signaling pathways that cooperate and form a complex network [62] (Figure 2).

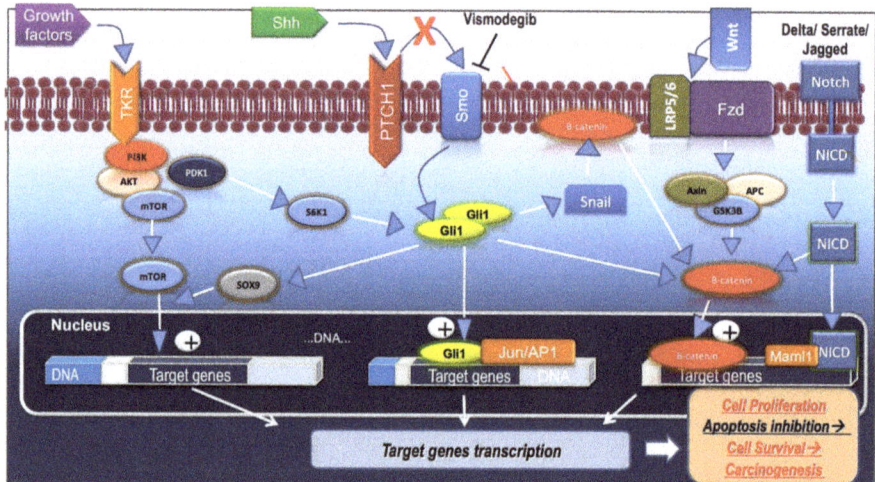

Figure 2. Crosstalk between Hh signaling pathway and PI3K/AKT/mTOR pathway, Wnt/β-catenin pathway, Notch pathway. Shh—Sonic hedgehog; PTCH1—protein patched homolog 1; SMO—smoothened protein; GLI—glioma-associated oncogenes; TKR—tyrosine kinase receptor; PDK-1—phosphoinositide-dependent kinase-1; PI3K—phosphoinositide 3-kinase; AKT—protein kinase B (PKB); mTOR—mammalian target of rapamycin; S6K1— S6 kinase 1; FZD—Frizzled receptors; LRP—low-density lipoprotein receptor related protein; APC—adenomatous polyposis coli; GSK3β—glycogen synthase kinase β; NICD—the intracellular fragment of Notch; Maml1—mastermind-like protein 1.

5.1. Wnt/β-Catenin Pathway

Wnt proteins are a complex of 19 lipidated and glycosylated proteins, which govern the activity of the canonical, β-catenin-dependent, and non-canonical, β-catenin-independent, Wnt pathways. In the non-canonical pathway, β-catenin does not undergo activation [63]. The Wnt/β-catenin pathway mediates numerous processes such as cell proliferation, migration and invasion and is involved in the development of several cancers. β-catenin is a member of a multi-molecular complex consisting of axin, adenomatous polyposis coli (APC) and glycogen synthase kinase β (GSK3β). In the absence of Wnt signaling, β-catenin is phosphorylated by GSK3β, ubiquitinated and subsequently it is degraded into the proteasome. When a Wnt ligand binds to the Frizzled receptors and the low-density lipoprotein receptor related protein (LRP), GSK3β is inactivated, β-catenin escapes from the complex and is translocated to the nucleus where Wnt target genes are upregulated [64,65]. Mutations in the Wnt pathway can lead to its activation independently of ligands and subsequently the malignant process is initiated. Under normal conditions, the Wnt pathway is inactive [65].

The Hh pathway downregulates the Wnt pathway through secreted frizzled-related protein 1 (SFRP1), and the Wnt pathway modulates the activity of the Hh pathway through GLI3. A disruption in this antagonism may be involved in tumorigenesis [66]. In BCC, the activation of the Hh pathway can induce aberrant activation of the Wnt pathway by the GLI transcription factors. The crosstalk between the two pathways is mediated by several molecules. Wnt2b, Wnt4, and Wnt7b are activated by GLI1 and subsequently the Wnt/β-catenin pathway is stimulated. In addition, β-catenin can increase the expression of the coding region determinant-binding protein (CRD-BP) and thus promotes the stabilization of GLI mRNAs [65]. Noubissi et al. showed that in BCC the expression of CRD-BP is increased and there is a positive correlation between CRD-BP level and the activation of Wnt and Hh signaling pathways. A decreased CRD-BP expression is linked to a low proliferation rate of BCC cells [67]. Alternatively, GLI1 simulates Wnt proteins and Snail to promote the translocation of β-catenin from the cell's membrane to its nucleus; in the cell membrane, β-catenin forms a complex with E-cadherin and Snail acts as a suppressor of E-cadherin (Figure 2) [68]. In about 30% of BCC samples, it has been found an accumulation of β-catenin in the nucleus, an accumulation that is associated with a higher proliferation rate [69]. GSK3β is not only a member of the Wnt pathway, but it has also been demonstrated that GSK3β participates in the Hh pathway as well, being involved in the activation of Snail by GLI1 [68]. In the skin, the Wnt pathway may also interact with the Notch and vitamin D pathways [63].

However, Carmo et al. have found a downregulation of Wnt3 and Wnt16 in 58 nodular BCC samples when compared to healthy tissue. Wnt3 can activate both canonical and non-canonical pathways and is involved in cell proliferation and malignant transformation. Wnt3 overexpression has been commonly identified in aggressive tumors. Carmo et al. pointed out that BCC is not an aggressive tumor, therefore there may be different gene expression profiles in such tumors [70].

Brinkhuizen et al. have shown that promoter hypermethylation of the components of the Hh and Wnt pathways is involved in carcinogenesis in BCC, a finding that may underlie the development of new therapies [12]. Interestingly, it appears that in some cases of BCC treated with Hh inhibitors, the Wnt pathway could play a role in relapse by modulating the transcriptional profile of the residual cells [69].

5.2. PI3K/AKT/mTOR Pathway

The PI3K family comprises enzymes with multiple subunits that act jointly to induce the conversion of phosphatidylinositol diphosphate (PIP2) to phosphatidylinositol triphosphate (PIP3). PIP3 via phosphoinositide-dependent kinase-1 (PDK-1), promotes the phosphorylation of AKT, a serine/threonine kinase, and its conversion to the active form. AKT can act on many targets, one of the main targets being mTOR. Other important targets are cyclic AMP–responsive element binding protein (CREB), and procaspase 9, p21, p27 families [71]. The PI3K/AKT signaling pathway is involved in mTOR phosphorylation. The downstream effector of PI3K is mTOR, which also acts as an upstream regulator [72].

The PI3K/AKT/mTOR pathway plays an important role in the normal growth and development of the human body. Mutations in the components of this pathway can lead to alterations that modify mTOR signaling, therefore the aberrant mTOR signaling pathway has been identified in various disorders [73,74]. mTOR mediates cell growth and alterations in mTOR pathway have been related to the development of several other neoplasms [72]. Certain growth factors and oncogenic proteins act as activators of the PI3K/AKT/mTOR pathway. The stimulation of PI3K/AKT/mTOR signaling promotes the phosphorylation and activation of several protein kinases, which are involved in carcinogenesis [62].

mTOR is a serine/threonine kinase that pertains to the PI3K-related protein kinase family; its C-terminus exhibits a great structural similarity to the catalytic domain of PI3K. mTOR includes two protein complexes, mTOR complex 1 (mTORC1), and mTOR complex 2 (mTORC2), that have different functions. mTORC1 is upregulated by the PI3K/AKT signaling and downregulated by the TSC1/TSC2 complex. The downstream targets of mTORC1 are S6K1 and 4EBP1, which control mRNA translation. mTORC2 is upregulated by growth factors, activates PKC-α and AKT and regulates the function of the small GTPases (Rhoa, Rac1 and Cdc42), involved in cell survival and modulation of the actin cytoskeleton [73,75].

Kim et al. highlighted that in BCC there is a crosstalk between Hh and PI3K/AKT/mTOR pathways. SOX9, a protein whose expression is mediated by GLI, stimulates mTOR transcriptional activity. Moreover, depletion of SOX9 is associated with a decreased mTOR expression and consequently a decreased BCC cell proliferation [76]. PI3K induces PDK1 activation which in turn will activate S6K1.

S6K1 can phosphorylate GLI1, thus GLI1-SUFU interaction is blocked, GLI1 is translocated to the nucleus inducing GLI-dependent transcription (Figure 2) [42].

The inhibition of mTOR is associated with the activation of another important cellular process, autophagy [43]. The latest studies point out that the role of autophagy in tumorigenesis should be studied more deeply as autophagy is interconnected with Hh signaling. Autophagy is an important process responsible for the elimination of damaged cells, being involved in tumor initiation and progression. It seems that Hh signaling has both a stimulatory and inhibitory effect on autophagy, but most studies have revealed its inhibitory role [34].

Everolimus, an immunosuppressive agent, acting on mTOR has shown encouraging results in BCC therapy [77]. The use of inhibitors of PI3K/AKT/mTOR pathway in combination with SMO inhibitors may enhance the effect of SMO inhibitors leading to a better response in BCC [78].

5.3. Hippo-YAP Pathway

The Hippo-YAP pathway mediates important cell processes such as cell differentiation, proliferation and apoptosis and through its downstream effectors, YAP and TAZ, is responsible for skin barrier function. In the damaged skin areas YAP and TAZ activate the stem cells involved in tissue regeneration [79]. YAP and TAZ play an essential role in embryonic development. At the same time, YAP and TAZ may contribute to carcinogenesis through the activation of target genes that promote cell proliferation, epithelial-to-mesenchymal transition (EMT) and metastasis [80].

Recent research has revealed that in BCC, the overexpression of the Hippo-YAP pathway participates in the process of tumorigenesis [79]. YAP and TAZ are two molecules that shuttle between the nucleus and the cytoplasm. In the nucleus, YAP and TAZ stimulate the expression of proliferative and antiapoptotic genes following the interaction with transcriptional factors of the TEA domain family members (TEAD). It has been observed that aberrant activation of the nuclear form of YAP is associated with basal cell proliferation and decreased markers of cell differentiation. However, the mechanism by which YAP initiates carcinogenesis in BCC is still unknown. Mutations in some genes, that control YAP and TAZ phosphorylation—*LATS1* and *LATS2* and their translocation from the nucleus to the cytoplasm—*PTN14*, are involved in the aberrant activation of the Hippo signaling [81]. The role of the Hippo pathway in BCC carcinogenesis is supported by the study conducted by Bonilla et al. which analyzed 293 BCC samples and showed that YAP target genes are overexpressed [14]. The study performed by Maglic et al. has found that Hippo signaling induces BCC carcinogenesis via the c-JUN/AP1 axis [82].

In a mouse model, Akladios et al. have revealed that positive regulatory interactions between YAP and Hh signaling are involved in BCC development. They showed that epidermal YAP activity induces the accumulation of GLI2 into nucleus in YAP2-5SA-ΔC mice [83].

5.4. EGFR Pathway

The epidermal growth factor receptor (EGFR) belongs to the ErbB family of tyrosine kinase receptors and stimulates the growth of cells previously activated by an EGFR ligand [84]. The specific ligands of EGFR are the epidermal growth factor, amphiregulin, TGF or heparin growth factor [85]. The binding of soluble ligands to the ectodomain of the receptor leads to homo and heterodimerization with other members of the receptor family. Receptor dimerization is a key step for the activation of its intracellular tyrosine kinase domain. Phosphotyrosine residues activate signaling pathways including Ras/MAPK, PLCγ1/PKC, PI3K/AKT, and STAT pathways [86]. EGFR overexpression is found in various tumors and represents an important promoter for the activation of different signaling pathways, leading to cell proliferation, invasion and metastasis [86]. EGFR is involved in some cases of SCC; 7% of head and neck SCC (HNSCC) display EGFR mutations [87].

Recent studies attribute a role to the EGFR pathway in BCC. Avci et al. detected a high EGFR expression in BCC samples, identifying a 4.17-fold increased expression in tumoral tissue compared to healthy tissue. In addition, the EGFR expression was 6.66 times higher in recurrent BCC compared with non-recurrent BCC. Analyzing the histopathological type, they concluded that EGFR can be considered a negative prognostic marker for infiltrative BCC with important consequences in terms of resection margins. The results were not statistically significant in the case of nodular and superficial BCC [88]. Similarly, another study found an increased EGFR expression in the analyzed BCC samples. The highest expression of EGFR was identified in the adenoid and morpheiform types and the lowest in the nodular type, suggesting that EGFR plays a role in the histological differentiation of BCC [85].

In vitro studies have shown that the interaction between Hh and EGFR pathways modulates the Hh target genes. The cooperation between EGFR and Hh signaling promotes the activation of RAS/MEK/ERK and JUN/AP-1 signaling (Figure 1). EGFR/Hh signaling is involved in the up-regulation of several genes required for BCC development including *SOX2, SOX9, JUN, CXCR4,* and *FGF19* [89]. Moreover, EGFR by activating ERK1/2 suppresses GLI2 proteolytic degradation in keratinocytes [43].

The study performed by Schnidar et al. emphasized the usefulness of a therapy based on the combined inhibition of the Hh and EGFR pathways. It has been observed that BCC cells express certain EGFR ligands, indicating the autocrine stimulation of this pathway [90]. Therapy with cetuximab, a monoclonal antibody that inhibits EGFR, has revealed promising results in keratinocyte carcinomas (BCC and SCC) [91].

5.5. Vitamin D Pathway

The action of vitamin D in cancer seems to be dual, with both pro- and anti-carcinogenic effects. The activation of vitamin D receptor (VDR) in the skin induces an antiproliferative effect by stimulating or inhibiting different pathways. [92,93]. In the skin, vitamin D inhibits the Hh signaling pathway as a protective mechanism against the harmful effects of UVB radiation. It has been observed that, in *Vdr*-null mice, the Hh pathway is overexpressed in the epidermis and hair follicle. Lack of VDR in keratinocytes interferes with cell differentiation, tissue repair, and increases the risk of developing a malignant process [94]. Teichert et al. have shown that vitamin D may directly inhibit the Hh pathway in a VDR-dependent manner. However, vitamin D might inhibit Hh pathway independently of VDR [95].

Recent studies have shown that vitamin D suppresses the Hh pathway by inhibiting SMO function. The mechanism is not fully understood but it has been shown that vitamin D acts upstream of PTCH and downstream of GLI. Another argument that vitamin D represses the Hh pathway by inhibiting SMO function is that there is no inhibition of the Hh pathway in the case of *SMO*-null cells [96]. Thus, Uhman et al. have shown that the application of calcitriol, the active form of vitamin D3, on the skin represses the development of BCC in *Ptch* mutant mice [97]. Calcitriol activates the VDR signaling pathway resulting in an antiproliferative effect and mediates cell differentiation by increasing the expression of markers such as involucrin, loricrin, and filaggrin. Moreover, calcitriol can mediate skin apoptosis [98]. Calcitriol is secreted by fibroblasts and released under the action of PTCH. It was pointed out that in *PTCH*-null cells, the synthesis of calcitriol occurs but the compound cannot be released [96].

However, the study by Brinkhuizen et al. did not reveal the efficacy of calcitriol in the treatment of superficial BCC. They have also tried a combination between diclofenac and calcitriol to reveal a synergistic effect, but there were no results. However, the use of diclofenac 3% gel in hyaluronic acid in BCC promotes apoptosis and inhibits cell proliferation [99].

5.6. P53 Pathway

There are several studies providing data on the role of *p53* in BCC pathogenesis. It has been shown that *p53* is overexpressed in BCC samples and suggested that *p53* mutations following chronic UV exposure might be an important factor in BCC development [100]. *P53* is a well-known tumor suppressor gene and has important implications for cancer prevention; therefore, mutations in the *p53* gene have been identified in various neoplasms. P53 protein may undergo inactivation by interacting with various proteins such as MDM-2, MDMx, and FAK [101]. It acts as a transcription factor by binding to certain sequences in the DNA structure leading to the activation or suppression of target genes. Thus, p53 controls pathways involved in cell division and DNA repair [102].

The role of *p53* mutations in BCC pathogenesis is not clear. In this regard, Oh et al. conducted the first study that showed an increased expression of p53, ΔNp63, TAp73, and γ-H2AX associated with the downregulation of MDM-2 [103]. The *p63* gene has two isoforms, TAp63 acts as a suppressor, whereas ΔNp63 acts as an oncogene. Similarly, the *p73* gene has two isoforms, TAp63, with tumor suppressor effect and ΔNp73 with an oncogenic role. Exposure to UVB leads to the production of γ-H2AX, which can be regarded as a marker of UVB-related DNA damage. MDM-2 is a negative modulator of p53 [103,104]. The study by Wang et al. tried to answer the question regarding the mechanism by which p53 is activated in BCC. The study has shown that aberrant Hh signaling activates p53 via Arf. In addition, the study has revealed that loss of p53 results in tumor development and progression. In contrast, loss of Arf is not associated with the initiation of the malignant process but is involved in tumor progression. On the other hand, an increased Arf expression in tumor keratinocytes contributes to the suppression of BCC development. The stress induced by oncogenes results in Arf activation which induces an increased p53 expression [105]. Arf/p53 pathway is involved in the elimination of altered cells [106].

The study by Li et al. has revealed that one of the mechanisms by which Hh pathway is involved in BCC tumorigenesis is the evasion of p53 activity. Moreover, Hh signaling contributes to p53 degradation in mouse embryonic fibroblasts [107].

Alterations of p53 function have implications for the treatment of BCC. A recent study on a BCC cell line has found that imiquimod promotes reactive oxygen species production, which will stimulate ATM and ATR signaling pathways contributing to cell apoptosis mediated by p53. In addition, cell lines that displayed mutations of p53 were more resistant to imiquimod-induced apoptosis [108].

5.7. Notch Pathway

The Notch signaling pathway could be involved in BCC tumorigenesis. Notch receptors are a group of four transmembrane proteins (Notch1-4) that are able to interact with different ligands. The most popular ligands are jagged 1 and 2 and delta 1, 3, and 4 [109]. The binding of the specific ligands promotes the activation of Notch signaling. This interaction produces the intramembrane cleavage of Notch receptor resulting in the release of the intracellular fragment of Notch (NICD), which translocates to the nucleus and activates the expression of Notch target genes [110].

Recent studies have shown that the interaction between Hh and Notch signaling pathways is involved in carcinogenesis and resistance to chemotherapy [111]. It has been found that Notch1 acts as an inhibitor of a malignant process. A study on animal models with Notch1-deficient skin has revealed a spontaneous development of BCC after a certain period. Moreover, in these cases, an increased activation of the Hh pathway was observed [109].

Shi et al. pointed out that there is a low expression of the Notch signaling pathway in BCC. After stimulating Notch signaling with Notch signaling peptide jagged 1, BCC cells undergo apoptosis. Interestingly, an increased activity of the Notch pathway was observed in the hair follicle, the origin of BCC. Thus, it was hypothesized that studying the Notch pathway in BCC may allow the introduction of new therapies [112]. Eberl et al. analyzed Notch expression in BCC cells and observed that the inner cells that display increased Notch activation after vismodegib treatment die, while those in the periphery that do not express Notch survive and lead to tumor recurrence. Thus, it seems that Notch modulation plays an important role in the pathogenesis and treatment of BCC [111].

Moreover, there is an important crosstalk between Notch and Wnt pathway. Wnt stimulates the expression of the Notch ligand Jagged, in turn Notch exerts an inhibitory effect on Wnt expression. In addition, mastermind-like protein 1 (Maml1), a coactivator of the Notch signaling, may act as a regulator of β-catenin transcription (Figure 2) [113].

6. Conclusions

The pathogenesis of BCC is very complex. *PTCH1* mutations play a crucial role in activating the Hh pathway; however, additional mutations that promote BCC carcinogenesis have been identified. Recent studies have shown that there is a significant cross-talk between Hh signaling pathway and other signaling pathways, including Wnt, Notch, EGFR, p53, PI3K/mTOR, and vitamin D. A further argument for the involvement of other pathways in the development of BCC could be the tumor resistance to Hh inhibitors. The knowledge and characterization of the BCC signaling pathways and the interactions between them could underlie the development of new therapies in BCC.

Author Contributions: All authors have equally contributed to this paper. Conceptualization, M.T., S.R.G. and M.N.; methodology, C.I.M. and M.I.M.; investigation, C.M. and C.S.; data curation C.C., C.S. and C.M.; writing—original draft preparation M.T., C.I.M. and M.I.M.; writing—review and editing, C.C., S.R.G. and M.N.; funding acquisition M.N. All authors have read and agreed to the published version of the manuscript.

Funding: This work was partially supported by 61PCCDI/2018 PN-III-P1-1.2-PCCDI-2017-0341—acronym PATHDERM and PN 19.29.01.01/2019.

Conflicts of Interest: The authors declare no conflict of interest.

References

1. Telfer, N.; Colver, G.; Morton, C. British Association of Dermatologists. Guidelines for the management of basal cell carcinoma. *Br. J. Dermatol.* **2008**, *159*, 35–48. [CrossRef] [PubMed]
2. Ghita, M.A.; Caruntu, C.; Rosca, A.E.; Kaleshi, H.; Caruntu, A.; Moraru, L.; Docea, A.O.; Zurac, S.; Boda, D.; Neagu, M.; et al. Reflectance confocal microscopy and dermoscopy for in vivo, non-invasive skin imaging of superficial basal cell carcinoma. *Oncol. Lett.* **2016**, *11*, 3019–3024. [CrossRef] [PubMed]
3. Tay, E.Y.-X.; Teoh, Y.-L.; Yeo, M.S.-W. Hedgehog Pathway Inhibitors and Their Utility in Basal Cell Carcinoma: A Comprehensive Review of Current Evidence. *Dermatol. Ther. Heidelb.* **2018**, *9*, 33–49. [CrossRef] [PubMed]
4. Situm, M.; Buljan, M.; Bulat, V.; Lugović-Mihić, L.; Bolanča, Z.; Šimić, D. The role of UV radiation in the development of basal cell carcinoma. *Coll. Antropol.* **2008**, *32*, 167–170.
5. Lupu, M.; Caruntu, C.; Popa, M.I.; Voiculescu, V.M.; Zurac, S.; Boda, D. Vascular patterns in basal cell carcinoma: Dermoscopic, confocal and histopathological perspectives (Review). *Oncol. Lett.* **2019**, *17*, 4112–4125. [CrossRef]
6. Crowson, A.N. Basal cell carcinoma: Biology, morphology and clinical implications. *Modern Pathol.* **2006**, *19*, S127–S147. [CrossRef]
7. Rubin, A.I.; Chen, E.H.; Ratner, D. Basal-Cell Carcinoma. *N. Engl. J. Med.* **2005**, *353*, 2262–2269. [CrossRef]
8. Durmaz, C.D.; Evans, G.; Smith, M.J.; Ertop, P.; Akay, B.N.; Tuncalı, T. A Novel PTCH1 Frameshift Mutation Leading to Nevoid Basal Cell Carcinoma Syndrome. *Cytogenet. Genome Res.* **2018**, *154*, 57–61. [CrossRef]
9. Almomani, R.; Khanfar, M.; Bodoor, K.; Al-Qarqaz, F.; Alqudah, M.; Hammouri, H.; Abu-Salah, A.; Haddad, Y.; Al Gargaz, W.; Mohaidat, Z. Evaluation of Patched-1 Protein Expression Level in Low Risk and High Risk Basal Cell Carcinoma Subtypes. *Asian Pac. J. Cancer Prev.* **2019**, *20*, 2851–2857. [CrossRef]
10. Muzio, L.L. Nevoid basal cell carcinoma syndrome (Gorlin syndrome). *Orphanet J. Rare Dis.* **2008**, *3*, 32. [CrossRef]
11. Narang, A.; Maheshwari, C.; Aggarwal, V.; Bansal, P.; Singh, P. Gorlin-Goltz Syndrome with Intracranial Meningioma: Case Report and Review of Literature. *World Neurosurg.* **2020**, *133*, 324–330. [CrossRef]
12. Brinkhuizen, T.; Hurk, V.D.K.; Winnepenninckx, V.J.L.; De Hoon, J.P.; Van Marion, A.M.; Veeck, J.; Van Engeland, M.; Van Steensel, M.A.M. Epigenetic Changes in Basal Cell Carcinoma Affect SHH and WNT Signaling Components. *PLoS ONE* **2012**, *7*, e51710. [CrossRef]
13. Montagna, E.; Lopes, O.S. Molecular Basis of Basal Cell Carcinoma. *An. Bras. Dermatol.* **2017**, *92*, 517–520. [CrossRef]
14. Bonilla, X.; Parmentier, L.; King, B.; Bezrukov, F.; Kaya, G.; Zoete, V.; Seplyarskiy, V.B.; Sharpe, H.J.; McKee, T.; Letourneau, A.; et al. Genomic Analysis Identifies New Drivers and Progression Pathways in Skin Basal Cell Carcinoma. *Nat. Genet.* **2016**, *48*, 398–406. [CrossRef]

15. Heller, E.R.; Gor, A.; Wang, D.; Hu, Q.; Lucchese, A.; Kanduc, D.; Katdare, M.; Liu, S.; Sinha, A.A. Molecular Signatures of Basal Cell Carcinoma Susceptibility and Pathogenesis: A Genomic Approach. *Int. J. Oncol.* **2013**, *42*, 583–596. [CrossRef]
16. Jayaraman, S.S.; Rayhan, D.J.; Hazany, S.; Kolodney, M.S. Mutational Landscape of Basal Cell Carcinomas by Whole-Exome Sequencing. *J. Investig. Dermatol.* **2014**, *134*, 213–220. [CrossRef]
17. Dai, J.; Lin, K.; Huang, Y.; Lu, Y.; Chen, W.-Q.; Zhang, X.-R.; He, B.-S.; Pan, Y.-Q.; Wang, S.-K.; Fan, W.-X. Identification of Critically Carcinogenesis-Related Genes in Basal Cell Carcinoma. *OncoTargets Ther.* **2018**, *11*, 6957–6967. [CrossRef]
18. Sharpe, H.J.; Pau, G.; Dijkgraaf, G.J.; Basset-Séguin, N.; Modrusan, Z.; Januario, T.; Tsui, V.; Durham, A.B.; Dlugosz, A.A.; Haverty, P.M.; et al. Genomic Analysis of Smoothened Inhibitor Resistance in Basal Cell Carcinoma. *Cancer Cell* **2015**, *27*, 327–341. [CrossRef]
19. Atwood, S.X.; Sarin, K.Y.; Whitson, R.J.; Li, J.R.; Kim, G.; Rezaee, M.; Ally, M.S.; Kim, J.; Yao, C.; Chang, A.L.S.; et al. Smoothened Variants Explain the Majority of Drug Resistance in Basal Cell Carcinoma. *Cancer Cell* **2015**, *27*, 342–353. [CrossRef]
20. Sekulic, A.; Migden, M.R.; Oro, A.E.; Dirix, L.; Lewis, K.D.; Hainsworth, J.D.; Solomon, J.A.; Yoo, S.; Arron, S.T.; Friedlander, P.A.; et al. Efficacy and Safety of Vismodegib in Advanced Basal-Cell Carcinoma. *N. Engl. J. Med.* **2012**, *366*, 2171–2179. [CrossRef]
21. Danial, C.; Sarin, K.Y.; Oro, A.E.; Chang, A.L.S. An Investigator-Initiated Open-Label Trial of Sonidegib in Advanced Basal Cell Carcinoma Patients Resistant to Vismodegib. *Clin. Cancer Res.* **2016**, *22*, 1325–1329. [CrossRef]
22. Kim, D.J.; Kim, J.; Spaunhurst, K.; Montoya, J.; Khodosh, R.; Chandra, K.; Fu, T.; Gilliam, A.; Molgó, M.; Beachy, P.A.; et al. Open-Label, Exploratory Phase II Trial of Oral Itraconazole for the Treatment of Basal Cell Carcinoma. *J. Clin. Oncol.* **2014**, *32*, 745–751. [CrossRef]
23. Siu, L.L.; Papadopoulos, K.; Alberts, S.R.; Kirchoff-Ross, R.; Vakkalagadda, B.; Lang, L.; Ahlers, L.M.; Bennett, K.L.; Van Tornout, J.M. A first-in-human, phase I study of an oral hedgehog (HH) pathway antagonist, BMS-833923 (XL139), in subjects with advanced or metastatic solid tumors. *J. Clin. Oncol.* **2019**, *28*, 2501. [CrossRef]
24. Bendell, J.C.; Andre, V.; Ho, A.; Kudchadkar, R.; Migden, M.; Infante, J.; Tiu, R.V.; Pitou, C.; Tucker, T.; Brail, L.; et al. Phase I Study of LY2940680, a Smo Antagonist, in Patients with Advanced Cancer Including Treatment-Naïve and Previously Treated Basal Cell Carcinoma. *Clin. Cancer Res.* **2018**, *24*, 2082–2091. [CrossRef]
25. Jimeno, A.; Weiss, G.J.; Miller, W.H.; Gettinger, S.; Eigl, B.J.C.; Chang, A.L.S.; Dunbar, J.; Devens, S.; Faia, K.; Skliris, G.; et al. Phase I Study of the Hedgehog Pathway Inhibitor IPI-926 in Adult Patients with Solid Tumors. *Clin. Cancer Res.* **2013**, *19*, 2766–2774. [CrossRef]
26. Peukert, S.; He, F.; Dai, M.; Zhang, R.; Sun, Y.; Miller-Moslin, K.; McEwan, M.; Lagu, B.; Wang, K.; Yusuff, N.; et al. Discovery of NVP-LEQ506, a Second-Generation Inhibitor of Smoothened. *Chem. Med. Chem.* **2013**, *8*, 1261–1265. [CrossRef]
27. Hassounah, N.B.; Bunch, T.A.; McDermott, K.M. Molecular Pathways: The Role of Primary Cilia in Cancer Progression and Therapeutics with a Focus on Hedgehog Signaling. *Clin. Cancer Res.* **2012**, *18*, 2429–2435. [CrossRef]
28. Lauth, M.; Bergström, A.; Shimokawa, T.; Toftgård, R. Inhibition of GLI-Mediated Transcription and Tumor Cell Growth by Small-Molecule Antagonists. *Proc. Natl. Acad. Sci. USA* **2007**, *104*, 8455–8460. [CrossRef]
29. Ally, M.S.; Ransohoff, K.; Sarin, K.Y.; Atwood, S.X.; Rezaee, M.; Bailey-Healy, I.; Kim, J.; Beachy, P.A.; Chang, A.L.S.; Oro, A.; et al. Effects of Combined Treatment With Arsenic Trioxide and Itraconazole in Patients With Refractory Metastatic Basal Cell Carcinoma. *JAMA Dermatol.* **2016**, *152*, 452–456. [CrossRef]
30. Jacobsen, A.A.; Aldahan, A.S.; Hughes, O.B.; Shah, V.V.; Strasswimmer, J. Hedgehog Pathway Inhibitor Therapy for Locally Advanced and Metastatic Basal Cell Carcinoma. A Systematic Review and Pooled Analysis of Interventional Studies. *JAMA Dermatol.* **2016**, *152*, 816–824. [CrossRef]
31. Basset-Séguin, N.; Sharpe, H.J.; De Sauvage, F.J. Efficacy of Hedgehog Pathway Inhibitors in Basal Cell Carcinoma. *Mol. Cancer Ther.* **2015**, *14*, 633–641. [CrossRef] [PubMed]
32. Whitson, R.J.; Lee, A.; Urman, N.M.; Mirza, A.; Yao, C.Y.; Brown, A.S.; Li, J.R.; Shankar, G.; Fry, M.A.; Atwood, S.X.; et al. Non-Canonical Hedgehog Pathway Activation by MKL1/SRF Promotes Drug-Resistance in Basal Cell Carcinomas. *Nat. Med.* **2018**, *24*, 271–281. [CrossRef]

33. Mahindroo, N.; Punchihewa, C.; Fujii, N. Hedgehog-Gli Signaling Pathway Inhibitors as Anticancer Agents. *J. Med. Chem.* **2009**, *52*, 3829–3845. [CrossRef]
34. Zeng, X.; Ju, D. Hedgehog Signaling Pathway and Autophagy in Cancer. *Int. J. Mol. Sci.* **2018**, *19*, 2279. [CrossRef]
35. Li, C.; Chi, S.; Xie, J. Hedgehog Signaling in Skin Cancers. *Cell. Signal.* **2011**, *23*, 1235–1243. [CrossRef]
36. Sari, I.N.; Phi, L.T.H.; Jun, N.; Wijaya, Y.T.; Lee, S.; Kwon, H.Y. Hedgehog Signaling in Cancer: A Prospective Therapeutic Target for Eradicating Cancer Stem Cells. *Cells* **2018**, *7*, 208. [CrossRef]
37. Gutzmer, R.; Solomon, J.A. Hedgehog Pathway Inhibition for the Treatment of Basal Cell Carcinoma. *Target. Oncol.* **2019**, *14*, 253–267. [CrossRef]
38. Brennan, D.; Chen, X.; Cheng, L.; Mahoney, M.; Riobo, N.A. Noncanonical Hedgehog Signaling. *Vitam. Horm.* **2012**, *88*, 55–72. [CrossRef]
39. Athar, M.; Li, C.; Kim, A.L.; Spiegelman, V.S.; Bickers, D.R. Sonic Hedgehog Signaling in Basal Cell Nevus Syndrome. *Cancer Res.* **2014**, *74*, 4967–4975. [CrossRef]
40. Tang, J.Y. Elucidating the Role of Molecular Signaling Pathways in the Tumorigenesis of Basal Cell Carcinoma. *Semin. Cutan. Med. Surg.* **2011**, *30*, S6–S9. [CrossRef]
41. Lupu, M.; Caruntu, C.; Ghita, M.A.; Voiculescu, V.; Voiculescu, S.; Rosca, A.E.; Caruntu, A.; Moraru, L.; Popa, I.M.; Calenic, B.; et al. Gene Expression and Proteome Analysis as Sources of Biomarkers in Basal Cell Carcinoma. *Dis. Markers* **2016**, *2016*, 1–9. [CrossRef] [PubMed]
42. Otsuka, A.; Levesque, M.P.; Dummer, R.; Kabashima, K. Hedgehog Signaling in Basal Cell Carcinoma. *J. Dermatol. Sci.* **2015**, *78*, 95–100. [CrossRef] [PubMed]
43. Bakshi, A.; Chaudhary, S.C.; Rana, M.; Elmets, C.A.; Athar, M. Basal Cell Carcinoma Pathogenesis and Therapy Involving Hedgehog Signaling and Beyond. *Mol. Carcinog.* **2017**, *56*, 2543–2557. [CrossRef]
44. Kim, H.S.; Kim, Y.S.; Lee, C.; Shin, M.S.; Kim, J.W.; Jang, B.G. Expression Profile of Sonic Hedgehog Signaling-Related Molecules in Basal Cell Carcinoma. *PLoS ONE* **2019**, *14*, e0225511. [CrossRef] [PubMed]
45. Teperino, R.; Aberger, F.; Esterbauer, H.; Riobó, N.; Pospisilik, J.A. Canonical and Non-Canonical Hedgehog Signalling and the Control of Metabolism. *Semin. Cell Dev. Biol.* **2014**, *33*, 81–92. [CrossRef] [PubMed]
46. Omland, S.H. Local Immune Response in Cutaneous Basal Cell Carcinoma. *Dan. Med. J.* **2017**, *64*, B5412.
47. Rangwala, S.; Tsai, K.Y. Roles of the Immune System in Skin Cancer. *Br. J. Dermatol.* **2011**, *165*, 953–965. [CrossRef]
48. Kaporis, H.G.; Guttman-Yassky, E.; Lowes, M.A.; Haider, A.S.; Fuentes-Duculan, J.; Darabi, K.; Whynot-Ertelt, J.; Khatcherian, A.; Cardinale, I.; Novitskaya, I.; et al. Human Basal Cell Carcinoma Is Associated with Foxp3+ T cells in a Th2 Dominant Microenvironment. *J. Investig. Dermatol.* **2007**, *127*, 2391–2398. [CrossRef]
49. Omland, S.H.; Nielsen, P.S.; Gjerdrum, L.M.R.; Gniadecki, R. Immunosuppressive Environment in Basal Cell Carcinoma: The Role of Regulatory T Cells. *Acta Derm. Venereol.* **2016**, *96*, 917–921. [CrossRef]
50. Georgescu, S.R.; Tampa, M.; Mitran, C.I.; Mitran, M.I.; Caruntu, C.; Caruntu, A.; Lupu, M.; Matei, C.; Constantin, C.; Neagu, M. Tumour Microenvironment in Skin Carcinogenesis. *Adv. Exp. Med. Biol.* **2020**, *1226*, 123–142. [CrossRef]
51. Nomikos, K.; Lampri, E.; Spyridonos, P.; Bassukas, I.D. Alterations in the Inflammatory Cells Infiltrating Basal Cell Carcinomas during Immunocryosurgery. *Arch. Dermatol. Res.* **2019**, *311*, 499–504. [CrossRef]
52. Hall, E.T.; Fernandez-Lopez, E.; Silk, A.W.; Dummer, R.; Bhatia, S. Immunologic Characteristics of Nonmelanoma Skin Cancers: Implications for Immunotherapy. *Am. Soc. Clin. Oncol. Educ. Book* **2020**, *40*, 398–407. [CrossRef]
53. Pellegrini, C.; Orlandi, A.; Costanza, G.; Di Stefani, A.; Piccioni, A.; Di Cesare, A.; Chiricozzi, A.; Ferlosio, A.; Peris, K.; Fargnoli, M.C. Expression of IL-23/Th17-Related Cytokines in Basal Cell Carcinoma and in the Response to Medical Treatments. *PLoS ONE* **2017**, *12*, e0183415. [CrossRef]
54. Grund-Groeschke, S.; Stockmaier, G.; Aberger, F. Hedgehog/GLI Signaling in Tumor Immunity—New Therapeutic Opportunities and Clinical Implications. *Cell Commun. Signal.* **2019**, *17*, 172. [CrossRef]
55. Otsuka, A.; Dreier, J.; Cheng, P.F.; Nägeli, M.; Lehmann, H.; Felderer, L.; Frew, I.J.; Matsushita, S.; Levesque, M.P.; Dummer, R. Hedgehog Pathway Inhibitors Promote Adaptive Immune Responses in Basal Cell Carcinoma. *Clin. Cancer Res.* **2015**, *21*, 1289–1297. [CrossRef]

56. Okada, F. Inflammation-Related Carcinogenesis: Current Findings in Epidemiological Trends, Causes and Mechanisms. *Yonago Acta Med.* **2014**, *57*, 65–72.
57. Georgescu, S.R.; Mitran, C.I.; Mitran, M.I.; Caruntu, C.; Sarbu, M.I.; Matei, C.; Nicolae, I.; Tocut, S.M.; Popa, M.I.; Tampa, M. New Insights in the Pathogenesis of HPV Infection and the Associated Carcinogenic Processes: The Role of Chronic Inflammation and Oxidative Stress. *J. Immunol. Res.* **2018**, *2018*, 5315816. [CrossRef]
58. Surcel, M.; Constantin, C.; Caruntu, C.; Zurac, S.; Neagu, M. Inflammatory Cytokine Pattern Is Sex-Dependent in Mouse Cutaneous Melanoma Experimental Model. *J. Immunol. Res.* **2017**, *2017*, 9212134. [CrossRef]
59. Multhoff, G.; Molls, M.; Radons, J. Chronic Inflammation in Cancer Development. *Front. Immunol.* **2012**, *2*, 98. [CrossRef]
60. Tampa, M.; Mitran, M.I.; Mitran, C.I.; Sarbu, I.; Matei, C.; Nicolae, I.; Caruntu, A.; Tocut, S.M.; Popa, M.I.; Caruntu, C.; et al. Mediators of Inflammation—A Potential Source of Biomarkers in Oral Squamous Cell Carcinoma. *J. Immunol. Res.* **2018**, *2018*, 1061780. [CrossRef]
61. Sternberg, C.; Gruber, W.; Eberl, M.; Tesanovic, S.; Stadler, M.; Elmer, D.P.; Schlederer, M.; Grund, S.; Roos, S.; Wolff, F.; et al. Synergistic Cross-Talk of Hedgehog and Interleukin-6 Signaling Drives Growth of Basal Cell Carcinoma. *Int. J. Cancer* **2018**, *143*, 2943–2954. [CrossRef]
62. Piérard-Franchimont, C.; Hermanns-Lê, T.; Paquet, P.; Herfs, M.; Delvenne, P.; Piérard, G.E. Hedgehog- and MTOR-Targeted Therapies for Advanced Basal Cell Carcinomas. *Future Oncol.* **2015**, *11*, 2997–3002. [CrossRef]
63. Li, J.; Ji, L.; Chen, J.; Zhang, W.; Ye, Z. Wnt/β-Catenin Signaling Pathway in Skin Carcinogenesis and Therapy. *Biomed. Res. Int.* **2015**, *2015*, 964842. [CrossRef]
64. Noubissi, F.K.; Yedjou, C.G.; Spiegelman, V.S.; Tchounwou, P.B. Cross-Talk between Wnt and Hh Signaling Pathways in the Pathology of Basal Cell Carcinoma. *Int. J. Environ. Res. Public Health* **2018**, *15*, 1442. [CrossRef]
65. Pelullo, M.; Zema, S.; Nardozza, F.; Checquolo, S.; Screpanti, I.; Bellavia, D. Wnt, Notch, and TGF-β Pathways Impinge on Hedgehog Signaling Complexity: An Open Window on Cancer. *Front. Genet.* **2019**, *10*, 711. [CrossRef]
66. Ding, M.; Wang, X. Antagonism between Hedgehog and Wnt Signaling Pathways Regulates Tumorigenicity (Review). *Oncol. Lett.* **2017**, *14*, 6327–6333. [CrossRef]
67. Noubissi, F.K.; Kim, T.; Kawahara, T.N.; Aughenbaugh, W.D.; Berg, E.; Longley, B.J.; Athar, M.; Spiegelman, V.S. Role of CRD-BP in the Growth of Human Basal Cell Carcinoma Cells. *J. Investig. Dermatol.* **2014**, *134*, 1718–1724. [CrossRef]
68. Li, X.; Deng, W.; Lobo-Ruppert, S.; Ruppert, J. Gli1 Acts through Snail and E-Cadherin to Promote Nuclear Signaling by β-Catenin. *Oncogene* **2007**, *26*, 4489–4498. [CrossRef]
69. Lang, C.M.R.; Chan, C.-K.; Veltri, A.; Lien, W.-H. Wnt Signaling Pathways in Keratinocyte Carcinomas. *Cancers Basel* **2019**, *11*, 1216. [CrossRef]
70. Carmo, N.G.D.; Sakamoto, L.H.T.; Pogue, R.; Mascarenhas, C.D.C.; Passos, S.K.; Felipe, M.S.S.; Andrade, R.V.D. Altered Expression of PRKX, WNT3 and WNT16 in Human Nodular Basal Cell Carcinoma. *Anticancer Res.* **2016**, *36*, 4545–4552. [CrossRef]
71. Gerson, S.L.; Caimi, P.F.; William, B.F.; Creger, R.J. Pharmacology and Molecular Mechanisms of Antineoplastic Agents for Hematologic Malignancies. *Hematology* **2018**, *57*, 849–912. [CrossRef]
72. Karayannopoulou, G.; Euvrard, S.; Kanitakis, J. Differential Expression of P-MTOR in Cutaneous Basal and Squamous Cell Carcinomas Likely Explains Their Different Response to MTOR Inhibitors in Organ-Transplant Recipients. *Anticancer Res.* **2013**, *33*, 3711–3714.
73. Pópulo, H.; Lopes, J.M.; Soares, P. The MTOR Signalling Pathway in Human Cancer. *Int. J. Mol. Sci.* **2012**, *13*, 1886–1918. [CrossRef]
74. Fruman, D.A.; Chiu, H.; Hopkins, B.D.; Bagrodia, S.; Cantley, L.C.; Abraham, R.T. The PI3K Pathway in Human Disease. *Cell* **2017**, *170*, 605–635. [CrossRef]
75. Chamcheu, J.C.; Roy, T.; Uddin, M.B.; Banang-Mbeumi, S.; Chamcheu, R.-C.N.; Walker, A.L.; Liu, Y.-Y.; Huang, S. Role and Therapeutic Targeting of the PI3K/Akt/mTOR Signaling Pathway in Skin Cancer: A Review of Current Status and Future Trends on Natural and Synthetic Agents Therapy. *Cells* **2019**, *8*, 803. [CrossRef]

76. Kim, A.L.; Back, J.H.; Chaudhary, S.C.; Zhu, Y.; Athar, M.; Bickers, D.R. SOX9 Transcriptionally Regulates mTOR-Induced Proliferation of Basal Cell Carcinomas. *J. Investig. Dermatol.* **2018**, *138*, 1716–1725. [CrossRef]
77. Eibenschutz, L.; Colombo, D.; Catricalà, C. Everolimus for Compassionate Use in Multiple Basal Cell Carcinomas. *Case Rep. Dermatol. Med.* **2013**, *2013*. [CrossRef]
78. Atwood, S.X.; Whitson, R.J.; Oro, A.E. Advanced Treatment for Basal Cell Carcinomas. *Cold Spring Harb. Perspect. Med.* **2014**, *4*. [CrossRef]
79. Miranda, M.M.; Lowry, W.E. Hip to the Game: YAP/TAZ Is Required for Nonmelanoma Skin Cancers. *EMBO J.* **2018**, *37*, e99921. [CrossRef]
80. Schlegelmilch, K.; Mohseni, M.; Kirak, O.; Pruszak, J.; Rodriguez, J.R.; Zhou, D.; Kreger, B.T.; Vasioukhin, V.; Avruch, J.; Brummelkamp, T.R.; et al. Yap1 Acts Downstream of α-Catenin to Control Epidermal Proliferation. *Cell* **2011**, *144*, 782–795. [CrossRef]
81. Debaugnies, M.; Sánchez-Danés, A.; Rorive, S.; Raphaël, M.; Liagre, M.; Parent, M.-A.; Brisebarre, A.; Salmon, I.; Blanpain, C. YAP and TAZ Are Essential for Basal and Squamous Cell Carcinoma Initiation. *EMBO Rep.* **2018**, *19*, e45809. [CrossRef]
82. Maglic, D.; Schlegelmilch, K.; Dost, A.F.; Panero, R.; Dill, M.T.; Calogero, R.A.; Camargo, F.D. YAP-TEAD Signaling Promotes Basal Cell Carcinoma Development via a c-JUN/AP1 Axis. *EMBO J.* **2018**, *37*, e98642. [CrossRef]
83. Akladios, B.; Reinoso, V.M.; Cain, J.E.; Wang, T.; Lambie, D.L.; Watkins, D.N.; Beverdam, A. Positive Regulatory Interactions between YAP and Hedgehog Signalling in Skin Homeostasis and BCC Development in Mouse Skin in Vivo. *PLoS ONE* **2017**, *12*, e0183178. [CrossRef]
84. Sasaki, T.; Hiroki, K.; Yamashita, Y. The Role of Epidermal Growth Factor Receptor in Cancer Metastasis and Microenvironment. *Biomed. Res. Int.* **2013**, *2013*, 546318. [CrossRef]
85. Florescu, D.E.; Stepan, A.E.; Mărgăritescu, C.; Ciurea, R.N.; Stepan, M.D.; Simionescu, C.E. The Involvement of EGFR, HER2 and HER3 in the Basal Cell Carcinomas Aggressiveness. *Rom. J. Morphol. Embryol.* **2018**, *59*, 479–484.
86. Wieduwilt, M.J.; Moasser, M.M. The Epidermal Growth Factor Receptor Family: Biology Driving Targeted Therapeutics. *Cell. Mol. Life Sci.* **2008**, *65*, 1566–1584. [CrossRef]
87. Willmore-Payne, C.; Holden, J.A.; Layfield, L.J. Detection of EGFR- and HER2-Activating Mutations in Squamous Cell Carcinoma Involving the Head and Neck. *Mod. Pathol.* **2006**, *19*, 634–640. [CrossRef]
88. Avci, C.B.; Kaya, I.; Ozturk, A.; Ay, N.P.O.; Sezgin, B.; Kurt, C.C.; Akyildiz, N.S.; Bozan, A.; Yaman, B.; Akalin, T.; et al. The Role of EGFR Overexpression on the Recurrence of Basal Cell Carcinomas with Positive Surgical Margins. *Gene* **2019**, *687*, 35–38. [CrossRef]
89. Eberl, M.; Klingler, S.; Mangelberger, D.; Loipetzberger, A.; Damhofer, H.; Zoidl, K.; Schnidar, H.; Hache, H.; Bauer, H.-C.; Solca, F.; et al. Hedgehog-EGFR Cooperation Response Genes Determine the Oncogenic Phenotype of Basal Cell Carcinoma and Tumour-Initiating Pancreatic Cancer Cells. *EMBO Mol. Med.* **2012**, *4*, 218–233. [CrossRef]
90. Schnidar, H.; Eberl, M.; Klingler, S.; Mangelberger, R.; Kasper, M.; Hauser-Kronberger, C.; Regl, G.; Kroismayr, R.; Moriggl, R.; Sibilia, M.; et al. Epidermal Growth Factor Receptor Signaling Synergizes with Hedgehog/GLI in Oncogenic Transformation via Activation of the MEK/ERK/JUN Pathway. *Cancer Res.* **2009**, *69*, 1284–1292. [CrossRef]
91. Scarpati, D.V.G.; Perri, F.; Pisconti, S.; Costa, G.; Ricciardiello, F.; Del Prete, S.; Napolitano, A.; Carraturo, M.; Mazzone, S.; Addeo, R. Concomitant Cetuximab and Radiation Therapy: A Possible Promising Strategy for Locally Advanced Inoperable Non-Melanoma Skin Carcinomas. *Mol. Clin. Oncol.* **2016**, *4*, 467–471. [CrossRef] [PubMed]
92. Reddy, K.K. Vitamin D Level and Basal Cell Carcinoma, Squamous Cell Carcinoma, and Melanoma Risk. *J. Investig. Dermatol.* **2013**, *133*, 589–592. [CrossRef] [PubMed]
93. Neagu, M.; Constantin, C.; Caruntu, C.; Dumitru, C.; Surcel, M.; Zurac, S. Inflammation: A Key Process in Skin Tumorigenesis. *Oncol. Lett.* **2019**, *17*, 4068–4084. [CrossRef] [PubMed]
94. Bikle, D.D.; Christakos, S. New Aspects of Vitamin D Metabolism and Action—Addressing the Skin as Source and Target. *Nat. Rev. Endocrinol.* **2020**, *16*, 234–252. [CrossRef] [PubMed]
95. Teichert, A.; Elalieh, H.; Elias, P.; Welsh, J.; Bikle, D.D. Over-Expression of Hedgehog Signaling Is Associated with Epidermal Tumor Formation in Vitamin D Receptor Null Mice. *J. Investig. Dermatol.* **2011**, *131*, 2289–2297. [CrossRef]

96. Linder, B.; Weber, S.; Dittmann, K.; Adamski, J.; Hahn, H.; Uhmann, A. A Functional and Putative Physiological Role of Calcitriol in Patched1/Smoothened Interaction. *J. Biol. Chem.* **2015**, *290*, 19614–19628. [CrossRef]
97. Uhmann, A.; Niemann, H.; Lammering, B.; Henkel, C.; Hess, I.; Nitzki, F.; Fritsch, A.; Prüfer, N.; Rosenberger, A.; Dullin, C.; et al. Antitumoral Effects of Calcitriol in Basal Cell Carcinomas Involve Inhibition of Hedgehog Signaling and Induction of Vitamin D Receptor Signaling and Differentiation. *Mol. Cancer Ther.* **2011**, *10*, 2179–2188. [CrossRef]
98. Albert, B.; Hahn, H. Interaction of Hedgehog and Vitamin D Signaling Pathways in Basal Cell Carcinomas. *Adv. Exp. Med. Biol.* **2014**, *810*, 329–341. [CrossRef]
99. Brinkhuizen, T.; Frencken, K.J.A.; Nelemans, P.J.; Hoff, M.L.S.; Kelleners-Smeets, N.W.J.; Hausen, A.Z.; Van Der Horst, M.P.J.; Rennspiess, D.; Winnepenninckx, V.J.L.; Van Steensel, M.A.M.; et al. The Effect of Topical Diclofenac 3% and Calcitriol 3 Mg/g on Superficial Basal Cell Carcinoma (SBCC) and Nodular Basal Cell Carcinoma (NBCC): A Phase II, Randomized Controlled Trial. *J. Am. Acad. Dermatol.* **2016**, *75*, 126–134. [CrossRef]
100. Shea, C.R.; McNutt, N.S.; Volkenandt, M.; Lugo, J.; Prioleau, P.G.; Albino, A.P. Overexpression of P53 Protein in Basal Cell Carcinomas of Human Skin. *Am. J. Pathol.* **1992**, *141*, 25–29.
101. Golubovskaya, V.M.; Cance, W.G. Targeting the P53 Pathway. *Surg. Oncol. Clin. N. Am.* **2013**, *22*, 747–764. [CrossRef]
102. Mandinova, A.; Lee, S.W. The p53 Pathway as a Target in Cancer Therapeutics: Obstacles and Promise. *Sci. Transl. Med.* **2011**, *3*, 64rv1. [CrossRef] [PubMed]
103. Oh, S.-T.; Stark, A.; Reichrath, J. The P53 Signalling Pathway in Cutaneous Basal Cell Carcinoma: An Immunohistochemical Description. *Acta Derm. Venereol.* **2020**, *100*, adv00098. [CrossRef]
104. Tampa, M.; Caruntu, C.; Mitran, M.; Mitran, C.; Sarbu, I.; Rusu, L.-C.; Matei, C.; Constantin, C.; Neagu, M.; Georgescu, S.-R. Markers of Oral Lichen Planus Malignant Transformation. *Dis. Markers* **2018**, *2018*, 1959506. [CrossRef]
105. Wang, G.Y.; Wood, C.N.; Dolorito, J.A.; Libove, E.; Epstein, E.H. Differing Tumor-Suppressor Functions of Arf and P53 in Murine Basal Cell Carcinoma Initiation and Progression. *Oncogene* **2017**, *36*, 3772–3780. [CrossRef]
106. Matheu, A.; Maraver, A.; Serrano, M. The Arf/p53 Pathway in Cancer and Aging. *Cancer Res.* **2008**, *68*, 6031–6034. [CrossRef]
107. Li, Z.J.; Mack, S.C.; Mak, T.H.; Angers, S.; Taylor, M.D.; Hui, C.-C. Evasion of P53 and G2/M Checkpoints Are Characteristic of Hh-Driven Basal Cell Carcinoma. *Oncogene* **2014**, *33*, 2674–2680. [CrossRef]
108. Huang, S.-W.; Chang, S.-H.; Mu, S.-W.; Jiang, H.-Y.; Wang, S.-T.; Kao, J.-K.; Huang, J.-L.; Wu, C.-Y.; Chen, Y.-J.; Shieh, J.-J. Imiquimod Activates P53-Dependent Apoptosis in a Human Basal Cell Carcinoma Cell Line. *J. Dermatol. Sci.* **2016**, *81*, 182–191. [CrossRef]
109. Nowell, C.; Radtke, F. Cutaneous Notch Signaling in Health and Disease. *Cold Spring Harb. Perspect. Med.* **2013**, *3*, a017772. [CrossRef]
110. D'Souza, B.; Meloty-Kapella, L.; Weinmaster, G. Canonical and Non-Canonical Notch Ligands. *Curr. Top. Dev. Biol.* **2010**, *92*, 73–129.
111. Eberl, M.; Mangelberger, R.; Swanson, J.B.; Verhaegen, M.E.; Harms, P.W.; Frohm, M.L.; Dlugosz, A.A.; Wong, S.Y. Tumor Architecture and Notch Signaling Modulate Drug Response in Basal Cell Carcinoma. *Cancer Cell* **2018**, *33*, 229–243. [CrossRef]
112. Shi, F.-T.; Yu, M.; Zloty, D.; Bell, R.H.; Wang, E.; Akhoundsadegh, N.; Leung, G.; Haegert, A.; Carr, N.; Shapiro, J.; et al. Notch Signaling Is Significantly Suppressed in Basal Cell Carcinomas and Activation Induces Basal Cell Carcinoma Cell Apoptosis. *Mol. Med. Rep.* **2017**, *15*, 1441–1454. [CrossRef]
113. Thompson, M.; Nejak-Bowen, K.; Monga, S.P.S. Crosstalk of the Wnt Signaling Pathway. In *Targeting the Wnt Pathway in Cancer*; Goss, K.H., Kahn, M., Eds.; Springer: New York, NY, USA, 2011; pp. 51–80. [CrossRef]

 © 2020 by the authors. Licensee MDPI, Basel, Switzerland. This article is an open access article distributed under the terms and conditions of the Creative Commons Attribution (CC BY) license (http://creativecommons.org/licenses/by/4.0/).

Article

Use of Cytology in the Diagnosis of Basal Cell Carcinoma Subtypes

Paola Pasquali [1,*], Gonzalo Segurado-Miravalles [2], Mar Castillo [3], Ángeles Fortuño [4], Susana Puig [3] and Salvador González [5]

1. Dermatology Department, Pius Hospital de Valls, 43800 Tarragona, Spain
2. Dermatology Department, Hospital Ramón y Cajal, 28034 Madrid, Spain; gonzalosegmi@hotmail.com
3. Dermatology Department, Hospital Clinic de Barcelona, IDIBAPS, University of Barcelona, Spain & CIBERER Barcelona, 08036 Barcelona, Spain; marcastillofort@gmail.com (M.C.); susipuig@gmail.com (S.P.)
4. Eldine Patología, 43006 Tarragona, Spain; afortunyo@eldinepatologia.org
5. Medicine and Medical Specialties Department, Alcalá University, 28801 Madrid, Spain; salvagonrod@gmail.com
* Correspondence: pasqualipaola@gmail.com; Tel.: +34-670-81-31-21

Received: 18 January 2020; Accepted: 18 February 2020; Published: 25 February 2020

Abstract: Background: Basal cell carcinoma (BCC) is the most common skin cancer in the white population. Nonsurgical treatments are first-line alternatives in superficial BCC (sBCC); therefore, differentiating between sBCC and non-sBCC is of major relevance for the clinician. Scraping cytology possesses several advantages, such as an earlier diagnosis and scarring absence, in comparison to a biopsy. Nevertheless, previous studies reported difficulties in differentiating the different BCC subtypes. The objective of this study was to determine the capability and accuracy of scraping cytology to differentiate between sBCC and non-sBCC. Methods: In this retrospective study, cytological samples of histologically confirmed BCC were examined. Select cytological features were correlated to BCC subtypes (sBCC or non-sBCC). Results: A total of 84 BCC samples were included (29 sBCC; 55 non-sBCC). An inverse correlation between the diagnosis of sBCC and the presence of mucin, dehiscence, and grade of atypia in the basal cells was observed. The presence of medium and large basal cell clusters correlated directly to a sBCC diagnosis. The presence of clear cells is strongly associated with sBCC. Therefore, Conclusion: Scraping cytology is reliable in differentiating sBCC from other BCC subtypes.

Keywords: skin cancer; diagnosis; cytology; basal cell carcinoma

1. Introduction

Basal cell carcinoma (BCC) is the most common skin cancer in the white population [1]. This slow-growing, malignant epithelial skin tumor predominantly affects older people; however, epidemiological data point out an increasing incidence, particularly in the younger population [2].

Although there is no global consensus on the classification of BCC subtypes, one of the most accepted divides them in a nodulocystic, adenoid, micronodular, infiltrative, morpheaform (sclerosing), keratotic, metatypical (basosquamous), pigmented, superficial, and ulcerative BCC. Other unusual variants are pleomorphic (giant cell), clear cell, signet ring cell, granular, infundibulocystic, metaplastic, shadow cell, and keloidal BCC [3].

Current guidelines for BCC management recommend a different approach depending on the BCC subtype. Thus, nonsurgical treatments are considered first-line treatments for superficial BCC (sBCC), whereas surgical alternatives are usually the first choice for other subtypes [4]. Therefore, BCC subtyping—or, at least, differentiating sBCC from non-superficial BCC (nsBCC)—is crucial for the clinician in order to choose surgical or nonsurgical treatments.

Currently, several non-invasive techniques, such as dermoscopy, high frequency ultrasound, and reflectance confocal microscopy, are used to identify the BCC subtype and support the treatment decision. However, histopathology remains the gold standard for BCC subtyping [5–8]. A skin biopsy is usually the technique performed for this purpose. The biopsy requires sample processing and it usually leaves a scar that, if the BCC is susceptible of nonsurgical treatment, would preferably be avoided, especially in cosmetically sensitive areas, such as the face.

Cytology is a non-commonly used technique in dermatology, unlike other specialties, such as gynecology. It has several advantages, such as earlier diagnosis, the absence of scars [9,10] and stitches, the sparing of local anesthesia and suture material, and it also saves the patient a trip back to an outpatients' minor procedures clinic to have the stitches removed [11].

Nevertheless, one of the most important drawbacks of exfoliative cytology so far is that previous studies reported that it is unable to differentiate the tumor subtypes [12], and others have only suggested its potential to determine the BCC subtype without proving it [13]. Scrape smears of BCC typically show many cohesive epithelial fragments composed of tightly-packed small cells with uniform, oval, dark nuclei. The nuclear chromatin is dense, but granular and evenly distributed; the nucleoli are small and indistinct. The cytoplasm is scanty and cyanophilic. The marginal palisading arrangement of tumor cells, stromal fragments, and mucin may be seen [9,14,15]. However, BCC subtyping—or, at least, differentiating sBCC from non-sBCC—would be of major relevance to the clinician in order to choose surgical or non-surgical management for BCC.

The objective of this study is to determine the capability and accuracy of cytology in differentiating sBCC from non-sBCC.

2. Material and Methods

A retrospective study was designed. Lesions with histological diagnosis of BCC and in which cytology had been previously done were included in the study. The lesions were collected between May 2016 and September 2016 from an outpatient clinic of the Pius Hospital of Valls and Eldine Laboratory (Tarragona, Spain). A previous informed consent was obtained from all patients.

Dermoscopy was performed prior the procedure to exclude coexisting lesions. The samples for cytology were obtained by a firm scraping of the lesion after first removing any surface crust. Usually, a scalpel blade was used. The tissue obtained was spread onto a glass slide and immediately fixed with fixation spray and stained using Papanicolaou's technique. Coverslips were placed on the slides with a Dibutylphthalate Polystyrene Xylene (DPX) mounting medium, a synthetic non-aqueous mounting medium for microscopy, and they were permanently filed. Cytological examination was done with a Leica DM750 microscope (Leica Microsystems, Wetzlar, Germany), and the image was taken with a Leica ICC50 camera (Leica, Wetzlar, Germany).

The cytological features were grouped into three broad categories: basal cells, squamous cells, and other findings. The system of evaluation is detailed in Table 1.

In this retrospective study, all the cytology smears were assessed blindly by a pathologist with 27 years of experience in cytology, with no cross-reference to the clinical notes or the routine histopathological report.

The biopsies were taken either by a shave biopsy or an excision following local anesthesia. They were fixed in 4% formaldehyde, routinely processed, and embedded in paraffin. Sections were stained with hematoxylin and eosin. Histopathological classification was based on the previously described standard criteria for each subtype, and included superficial, nodulocystic, and infiltrative BCC. Histopathologically, in the nsBCC group, nodulocystic and infiltrative BCCs were included.

Demographic data, such as age, sex, and anatomic location of the lesions were also collected.

Table 1. Cytological features of skin tissue scoring system definition.

Item		Definition		Score/Code
Basal cells	Cellularity	Total presence of basal cell clusters (plaques and/or groups) in the two extensions performed in each case (homogeneous behavior)	None Poor: <20 Moderate: 20–100 High: >100	0 1 2 3
	Groups	Three-dimensional cell clusters	Absence Presence	0 1
	Sheets	Two-dimensional cell clusters	Absence Presence	0 1
	Size	Size of cell clusters observed counted over 50 clusters in random fields. The percentages of the three groups add up 100%	Large clusters: >100 cells Medium clusters: 20–100 cells Small clusters: <20 cells	0–100 0–100 0–100
	Dehiscence	Quality of the peripheral cells of the cluster to be released from the primary cluster and remained isolated	Absence Presence	0 1
	Atypia	Abnormality in cells	Mild Moderate Severe	1 2 3
Squamous cells	Cellularity	Presence of squamous cellularity in the sample counted on 10HPF *	None Poor: <3 Moderate: 4–6 High: >6	0 1 2 3
Isolated cells		Quantity of dispersed single cells in the sample counted on 10HPF	0 <3 4–6 >6	0 1 2 3
Clustered cells		Quantity of clustered cells in the sample counted on 10HPF	None <3 4–6 >6	0 1 2 3
Other findings	Palisade	Peripheral cells of the cluster disposed as palisade cells	Absence Presence	0 1
	Mucin		Absence Presence	0 1
	Stroma	Presence of stromal fragments (Fibrous, fibromyxoid, fibrovascular) loose or attached to other cell groups. Valued over the entire sample	None <5 groups 6–10 groups >10 groups	0 1 2 3
	Clear cells	Presence of clear sebaceous cells	Absence Presence	0 1

* High power field.

Statistical analysis was performed using a XLSTAT statistical package (version 2.01.16684, 2015), considering each BCC as an independent event. The results were expressed as mean and standard deviation and frequencies. The outcome dichotomous variable was set to the definite histopathological diagnosis of a superficial type of BCC or a non-superficial type of BCC (including nodular and infiltrative types). All separated cytological variables were included in the analysis. On the one hand, the analysis of variance (ANOVA) and Chi-Square test were used to compare univariate associations of cytological features with a diagnosis of sBCC or non-sBCC. On the other hand, multivariate associations were assessed by using discriminant analysis (multiple logistic regression model), with the purpose of identifying independently significant cytological criteria to define each BCC subtype. All p-values cited are two-sided and p-values less than 0.05 were considered statistically significant. The accuracy of cytology sensitivity, specificity, positive predictive value, and negative predictive value of cytology for the sBCC diagnosis were calculated by comparing the cytological results with the histopathological findings. The results were arranged in a 2×2 contingency table.

3. Results

A total of 84 BCCs were included in the study from 45 patients (38 men, 84.4%, and 7 women, 15.6%). The age ranged from 52 to 96 (mean 76.5). The most common location was in the head and neck ($n = 52$), followed by the anterior and posterior thorax ($n = 28$), and extremities ($n = 4$). The major size of the tumors ranged from 3.77 mm to 13.00 mm (mean 8.35 mm, 95% CI = 7.7–9.0 mm).

The cytological findings relevant to the sBCC and nsBCC groups are summarized in Table 2.

Table 2. Cytological findings and distribution scores in superficial and non-superficial basal cell carcinoma (BCC).

Cytological Feature	Category	Superficial BCC (n = 29) Mean (95% CI)	Frequency	Non-Superficial BCC (n = 55) Mean (95% CI)	Frequency	p Value
Basal cells						
Cellularity	None Poor Moderate High	2.38 (2.09–2.67)	0% 3.4% 55.2% 41.4%	2.64 (2.34–2.94)	0% 10.9% 14.5% 74.6%	0.08
Groups	Absence Presence	1.00 (0.70–1.30)	0% 100%	1.34 (1.04–1.64)	1.8% 98.2%	0.02
Sheets	Absence Presence	0.62 (0.40–0.84)	37.9% 62.1%	0.60 (0.38–0.82)	40% 60%	0.86
Size	Large clusters Medium clusters Small clusters	33.8 (23.7–43.9) 34.3 (28.4–40.2) 31.9 (22.3–41.5)		46.8 (36.7–56.9) 22.7 (16.8–28.6) 30.45 (20.8–40.1)		0.01 0.01 0.77
Dehiscence	Absence Presence	0.10 (−0.09–0.29)	89.7% 11.3%	0.34 (0.15–0.53)	65.5% 34.5%	0.02
Atypia	Mild Moderate Severe	1.14 (0.87–1.41)	86.2% 13.8% 0%	1.73 (1.46–2.00)	36.4% 52.7% 10.9%	<0.0001
Squamous cells						
Cellularity	None Poor Moderate High	1.66 (1.17–2.15)	17.2% 20.7% 41.4% 20.7%	1.38 (0.89–1.87)	25.5% 32.7% 20% 21.8%	0.17
Isolated cells	0 <3 4–6 >6	2.48 (1.88–3.08)	17.2% 0% 0% 82.8%	2.02 (1.42–2.62)	29.1% 5.5% 0% 64.4%	0.13
Clustered cells	None <3 4–6 >6	1.52 (0.95–2.09)	34.5% 13.8% 17.2% 34.5%	0.96 (0.39–1.53)	50.9% 25.5% 0% 23.6%	0.04
Other findings						
Palisade	Absence Presence	0.69 (0.46–0.92)	31% 69%	0.56 (0.33–0.79)	43.6% 56.4%	0.27
Mucin	Absence Presence	0.07 (−0.11–0.25)	93.1% 6.9%	0.25 (0.07–0.43)	74.5% 25.5%	0.04
Stroma	None <5 groups 6–10 groups >10 groups	1.62 (1.12–2.12)	27.6% 20.7% 13.8% 37.9%	1.42 (0.92–1.92)	18.2% 38.2% 27.3% 16.3%	0.37
Clear cells	Absence Presence	0.45 (0.30–0.60)	55.2% 44.8%	0.02 (−0.13–0.17)	98.2% 1.8%	<0.0001

The results of the subtype BCC cytodiagnosis (Table 3) allowed us to calculate the accuracy of this technique. The sensitivity and specificity for BCC differentiation was 96.55% and 100%, whereas the positive and negative predictive values were 100% and 98.21%, respectively.

Table 3. Evaluation of cytology in subtype BCC diagnosis.

	Histopathology	
Cytology	Superficial BCC	Non-Superficial BCC
Superficial BCC	29	0
Non-superficial BCC	1	55

A multiple-group discriminant analysis, which included 18 parameters, correctly classified 75% of original grouped cases into superficial or non-superficial BCCs.

3.1. Superficial BCC Features

Upon cytological evaluation, the most frequent criteria of sBCC were moderate cellularity of the basal cells (16/29, 55.17%) forming groups with an equal distribution as large, medium, and small clusters (33.8%, 34.3%, and 31.9%, respectively, proportions), the presence of basal cell sheets (18/29, 62.07%), mild grade of basal cell atypia (25/29, 86.21%), and the absence of dehiscence (26/29, 88.67%).

In regards to the squamous cells visualized in the cytological analysis, the predominant pattern was a moderate grade of squamous cellularity (12/29, 41.38%), with a high proportion of isolated cells (24/29, 82.76%). Other common findings were the presence of palisade cells (20/29, 68.97%) and clear sebaceous cells (13/29, 44.83%), the absence of mucin (27/29, 93.10%), and finally, more than 10 groups of stromal fragments (11/29, 37.93%) (Figure 1A).

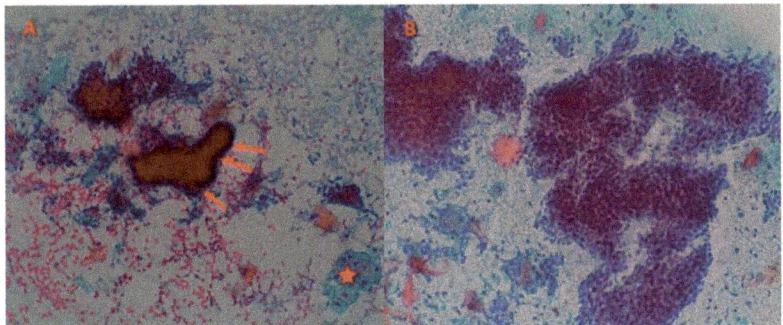

Figure 1. Basal cell cacinomas (BCCs). (**A**) Superficial basal cell carcinoma (sBCC): presence of two fragments composed of tightly packed small cells with mild atypia, peripheral palisade (red arrows), absence of dehiscence, and one cluster of clear sebaceous cells (red star) (Papanicolaou stain, ×100). (**B**) Non-superficial basal cell carcinoma (nsBCC): cellular smear with large clusters of basal cells with dehiscence and severe grade of atypia (Papanicolaou stain, ×200).

As far as anatomical sites go, the most common location of sBCC was the trunk (22/29, 75.86%) lesions, and head and neck for the rest of the superficial tumors (7/29, 24.14%).

According to the results and the statistics applied, the analyses have shown a strong inverse correlation between the diagnosis of sBCC and the presence of mucin, dehiscence, and atypia in basal cells (the more severe the atypia, the less likely of it being a sBCC). Moreover, a significantly direct correlation between the presence of medium and large basal cell clusters has also been discovered on this BCC subtype.

3.2. Non-Superficial BCC Features

Upon cytological examination, the nsBCCs revealed high cellularity (41/55, 74.55%) and groups and sheets of basal cells (54/55, 98.18% and 33/55, 60.00%, respectively). According to the frequency of visualization, the order of the basal cell cluster size was large clusters (46.81%), followed by small (30.45%), and finally, medium (22.69%). In comparison to the superficial subtype, 34.55% (19/55) of nsBCCs showed the presence of dehiscence. Moreover, 10.91% (6/55) of the samples had basal cells with a severe grade of atypia, although a moderate grade was the most common (29/55, 52.73%) (Figure 1B).

Among the cytological features analyzed in the squamous cells, poor cellularity was the most common pattern found (18/55, 32.73%), followed by a total absence of such cells in 25.5% of nsBCCs. Finally, in 65.45% (36/55) of the samples in this group, a high number of isolated cells was reported. In contrast, the cytology did not show any clustered cells in 50.91% (28/55) of cases. Out of the last four items evaluated in the cytology, the palisade image was present in 56.36% (31/55) of the tumors. The results frequently showed the absence of mucin and clear cells (41/55, 74.55%, and 54/55, 98.18%, respectively). Finally, the most viewed stroma form was as <5 groups of stromal fragments (21/55, 38.18%) (Figure 2).

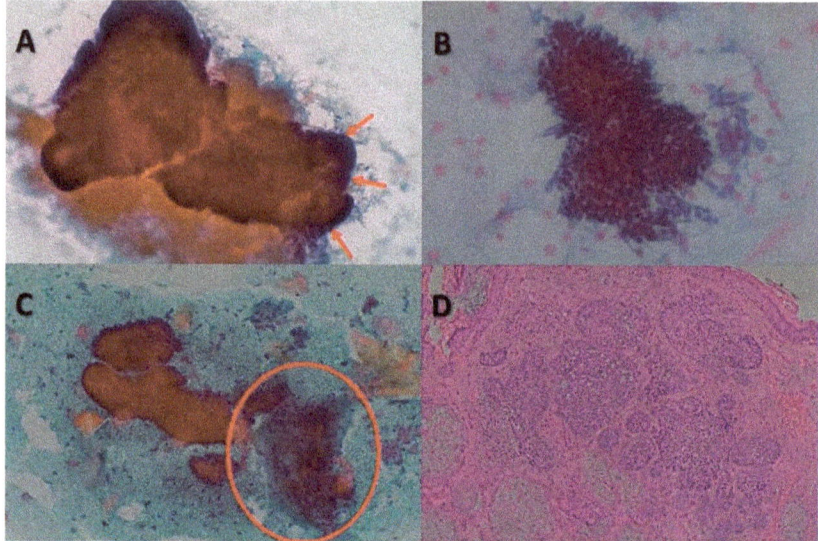

Figure 2. BCCs. (**A**) Cytology revealed large clusters of basal cells, mucin, and peripheral cells arranged in palisades (red arrows) (Papanicolaou stain, ×100). (**B**) Cytological image with a fragment composed of small basal cells with uniform oval, dark nuclei (Papanicolaou stain, ×200). (**C**) Group of basal cells accompanied by a stromal fragment (red circle) and keratinized squamous cells (Papanicolaou stain, ×100). (**D**) On histology, BCC showed a lobular pattern with islands and basaloid cells (hematoxylin and eosin, ×40).

Most commonly, the nsBCCs developed on the head and neck (45/55, 81.82%); whereas their presence on the trunk and extremities was only 10.91% (6/55) and 7.27% (4/55), respectively. Regarding the subtypes, most nsBCCs were nodular (35/55, 63.63%), followed by infiltrative (14/55, 25.45%), and other subtypes (6/55, 10.90%).

4. Discussion

This study is unique in evaluating the reliability of the cutaneous tissue smear cytodiagnosis of sBCC, independent of the histopathological evaluation. We assessed the accuracy of the cytological criteria for differentiating sBCC from other subtypes.

The clinical and epidemiological characteristics of sBCC in our study were comparable with existing evidence. As previously reported, sBCC was most commonly located on the trunk, whereas nsBCCs developed more frequently on the head and neck area [16].

Our results demonstrate that both classic and additional criteria may be present in superficial and non-superficial tumors, and accordingly, the presence of a single criterion cannot accurately predict the histopathological subtype. The clear cells represent an exception, being strongly associated with sBCC (Figure 3). In the only case where clear cells were found and nsBCC was established as the histopathological diagnosis, a biopsy also showed several focuses of sBCC. Furthermore, the frequent appearance of fragments containing a cluster of clear cells attached to a cluster of basal cells and stromal fragment could be explained by the superficiality of sBCC tumor proliferation and the intimate anatomical disposition of these three elements.

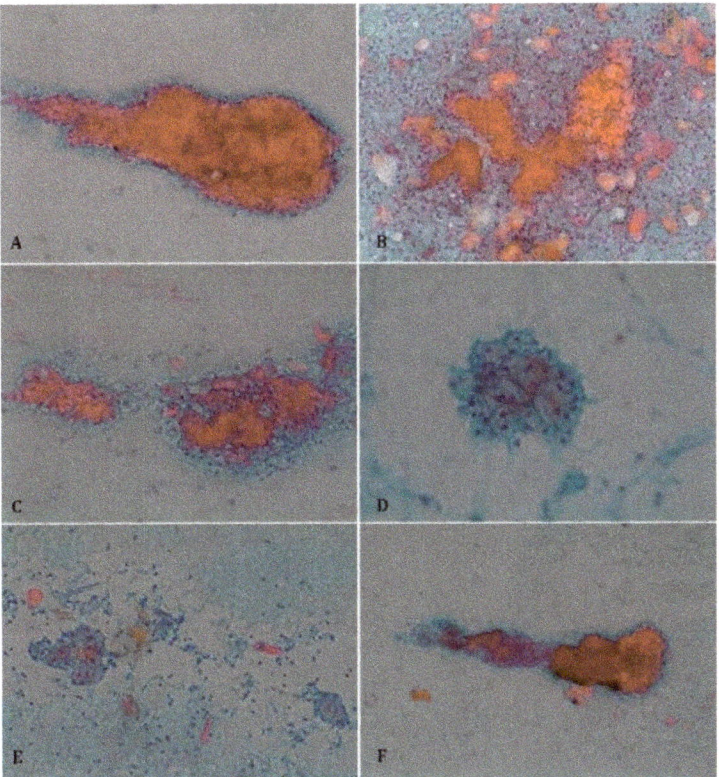

Figure 3. Clear sebaceous cells in sBCC (Papanicolaou stain). (**A**) Papanicolaou stain, ×100; (**B**) Papanicolaou stain, ×100; (**C**) Papanicolaou stain, ×100; (**D**) Papanicolaou stain, ×200; (**E**) Papanicolaou stain, ×100; (**F**) Papanicolaou stain, ×40.

According to our results, the cytological pattern of sBCC is formed by the low cellularity of basal cells in contrast to a higher amount of squamous cells. Regarding the basal cells, they are characterized by mild atypia with an absence of dehiscence. In addition, the presence of a high quantity of stroma, clear cells, and peripheral palisade was shown to predict a sBCC subtype. In contrast, the detection of mucin could be suggestive of excluding the diagnosis of sBCC. All these findings may be explained by several reasons. On the one hand, low basal cellularity, mild-grade atypia, the absence of dehiscence, and peripheral palisade are related to the well-differentiation of the BCC. On the other hand, the high content of squamous cells and a high amount of stroma are associated with the superficiality of tumor proliferation.

We have shown that cytology could be a reliable diagnostic method for differentiating BCC between sBCC and nsBCC, with 98.81% of the lesions being correctly assessed. The only false negative reported was an ulcerative BCC instead of a sBCC with focal ulceration. It is common to establish an ulcerative BCC misdiagnosis, since this variant is usually accompanied by a lot of granulation tissue and inflammation. Although there are other techniques, such as reflectance confocal microscopy and optical coherence tomography, that may be used for the same purpose, their higher costs may limit the access to numerous dermatologists.

Limitations of the present study include the retrospective design that is subject to recall and observer biases. In addition, no investigator has previously noted the cytological variability of different subtypes of BCCs. During the analysis of the first samples, there were no cytological variables known as significant to differentiate each BCC subtype, and actually, it was not certainly known if any cytological criteria would be found. Moreover, since the pathologist assessed the cytology smears blindly with no knowledge of the clinical notes or the final histopathological diagnosis, an association of cytological patterns with BCC subtypes could not be made in advance. Moreover, the shaving biopsy may have missed some of the deepest BCC subtypes in the patients it was performed on [17,18]. Although dermoscopy was performed to exclude coexisting lesions, the location of most BCCs over photodamaged skin of the head and neck may hamper the recording of the squamous cell due to the background epidermal atypia [19].

Related to the technical procedures, there were some difficulties in subtyping BCC with cytodiagnosis. Mucin visualization was difficult in some cytological smears due to two reasons: one of them is explained by the presence of a blood background in the evaluation of the samples, because of the cytological technique itself (scraping); the second reason lies in the staining technique. While using Diff-Quick, the mucin is observed with an intense fuchsia tonality; in Papanicolaou's staining, its intensity decreases significantly and the color is blue.

5. Conclusions

Cytological examination is easy to perform, does not require local anesthesia, saves time, is less expensive than a regular biopsy, and provides rapid diagnosis. Smear-taking for cytology is very well-tolerated, as it causes negligible trauma or discomfort to the patient.

Therefore, it can be performed (and, when necessary, repeated) even in apprehensive patients, and in sites where a biopsy has been proven to be difficult to obtain, or where aesthetic problems may arise, such as the face.

Our study revealed significant differences in the cytological characteristics among superficial and other BCC subtypes, suggesting that a combination of this technique with other non-aggressive diagnostic modalities, such as a clinical examination, dermatoscopy, or ultrasound, may significantly enhance the preoperative subtype classification of the tumor. This is particularly relevant in clinical practice, where treatment modality is determined by the tumor subtype.

Additional studies are needed to investigate whether cytology could increase the accuracy of the preoperative subtype classification of BCC.

Author Contributions: Conceptualization, P.P.; formal analysis, G.S.-M. and M.C.; funding acquisition, Á.F.; investigation, P.P. and Á.F.; methodology, P.P., M.C., Á.F., S.P. and S.G.; project administration, P.P. and Á.F.; supervision, P.P. and S.G.; visualization, G.S.-M., S.P. and S.G.; writing—original draft, M.C.; writing—review and editing, G.S.-M. and S.P. All authors have read and agreed to the published version of the manuscript.

Funding: This research was funded by Pius Hospital of Valls.

Conflicts of Interest: The authors declare that the research was conducted in the absence of any commercial or financial relationships that could a potential conflict of interest.

References

1. U.S. Department of Health and Human Services. Incidence of Nonmelanoma Skin Cancer in the United States. Available online: http://www.ciesin.org/docs/001-526/001-526.html (accessed on 31 October 2017).
2. Christenson, L.J.; Borrowman, T.A.; Vachon, C.M.; Tollefson, M.M.; Otley, C.C.; Weaver, A.L.; Roenigk, R.K. Incidence of basal cell and squamous cell carcinomas in a population younger than 40 years. *JAMA* **2005**, *294*, 681–690. [CrossRef] [PubMed]
3. Lever, W.F.; Elder, D.E. *Lever's Histopathology of the Skin*, 10th ed.; Lippincott Williams & Wilkins: Philadelphia, PA, USA, 2009.
4. Trakatelli, M.; Morton, C.; Nagore, E.; Ulrich, C.; Del Marmol, V.; Peris, K.; Basset-Seguin, N. Update of the European guidelines for basal cell carcinoma management. *Eur. J. Dermatol.* **2014**, *24*, 312–329. [CrossRef] [PubMed]

5. Longo, C.; Borsari, S.; Benati, E.; Moscarella, E.; Alfano, R.; Argenziano, G. Dermoscopy and Reflectance Confocal Microscopy for Monitoring the Treatment of Actinic Keratosis with Ingenol Mebutate Gel: Report of Two Cases. *Dermatol. Ther. (Heidelb)* **2016**, *6*, 81–87. [CrossRef] [PubMed]
6. Altamura, D.; Menzies, S.W.; Argenziano, G.; Zalaudek, I.; Soyer, H.P.; Sera, F.; Avramidis, M.; DeAmbrosis, K.; Fargnoli, M.C.; Peris, K. Dermatoscopy of basal cell carcinoma: Morphologic variability of global and local features and accuracy of diagnosis. *J. Am. Acad. Dermatol.* **2010**, *62*, 67–75. [CrossRef] [PubMed]
7. Yélamos, O.; Braun, R.P.; Liopyris, K.; Wolner, Z.J.; Kerl, K.; Gerami, P.; Marghoob, A.A. Dermoscopy and dermatopathology correlates of cutaneous neoplasms. *J. Am. Acad. Dermatol.* **2019**, *80*, 341–363. [CrossRef] [PubMed]
8. Villarreal-Martinez, A.; Bennàssar, A.; Gonzalez, S.; Malvehy, J.; Puig, S. Application of in vivo reflectance confocal microscopy and ex vivo fluorescence confocal microscopy in the most common subtypes of basal cell carcinoma and correlation with histopathology. *Br. J. Dermatol.* **2018**, *178*, 1215–1217. [CrossRef] [PubMed]
9. Fortuño-Mar, A. Cytology. In *Skin Cancer: A Practical Approach (Current Clinical Pathology)*; Springer: New York, NY, USA; Heidelberg, Germany; Dordrecht, The Netherlands; London, UK, 2014; pp. 213–219.
10. Oram, Y.; Turhan, O.; Aydin, N.E. Diagnostic value of cytology in basal cell and squamous cell carcinomas. *Int. J. Dermatol.* **1997**, *36*, 156–157. [CrossRef] [PubMed]
11. Bakis, S.; Irwig, L.; Wood, G.; Wong, D. Exfoliative cytology as a diagnostic test for basal cell carcinoma: A meta-analysis. *Br. J. Dermatol.* **2004**, *150*, 829–836. [CrossRef] [PubMed]
12. Vega-Memije, E.; De Larios, N.M.; Waxtein, L.M.; Dominguez-Soto, L. Cytodiagnosis of cutaneous basal and squamous cell carcinoma. *Int. J. Dermatol.* **2000**, *39*, 116–120. [CrossRef] [PubMed]
13. Gordon, L.A.; Orell, S.R. Evaluation of cytodiagnosis of cutaneous basal cell carcinoma. *J. Am. Acad. Dermatol.* **1984**, *11*, 1082–1086. [CrossRef]
14. Brown, C.L.; Klaber, M.R.; Robertson, M.G. Rapid cytological diagnosis of basal cell carcinoma of the skin. *J. Clin. Pathol.* **1979**, *32*, 361–367. [CrossRef] [PubMed]
15. Rege, J.; Shet, T. Aspiration cytology in the diagnosis of primary tumors of skin adnexa. *Acta Cytol.* **2001**, *45*, 715–722. [CrossRef] [PubMed]
16. McCormack, C.J.; Kelly, J.W.; Dorevitch, A.P. Differences in age and body site distribution of the histological subtypes of basal cell carcinoma. A possible indicator of differing causes. *Arch. Dermatol.* **1997**, *133*, 593–596. [CrossRef] [PubMed]
17. Pyne, J.H.; Myint, E.; Barr, E.M.; Clark, S.P.; David, M.; Na, R.; Hou, R. Superficial basal cell carcinoma: A comparison of superficial only subtype with superficial combined with other subtypes by age, sex and anatomic site in 3150 cases. *J. Cutan. Pathol.* **2017**, *44*, 677–683. [CrossRef] [PubMed]
18. Pyne, J.H.; Myint, E.; Barr, E.M.; Clark, S.P.; Hou, R. Basal cell carcinoma: Variation in invasion depth by subtype, sex, and anatomic site in 4,565 cases. *Dermatol. Pract. Concept.* **2018**, *8*, 314–319. [CrossRef] [PubMed]
19. Segurado-Miravalles, G.; Jiménez-Gómez, N.; Muñoz Moreno-Arrones, O.; Alarcón-Salazar, I.; Alegre-Sánchez, A.; Saceda-Corralo, D.; Jaén-Olasolo, P.; González-Rodríguez, S. Assessment of the Effect of 3% Diclofenac Sodium on Photodamaged Skin by Means of Reflectance Confocal Microscopy. *Acta Derm. Venereol.* **2018**, *98*, 963–969. [CrossRef] [PubMed]

© 2020 by the authors. Licensee MDPI, Basel, Switzerland. This article is an open access article distributed under the terms and conditions of the Creative Commons Attribution (CC BY) license (http://creativecommons.org/licenses/by/4.0/).

Review

A Systematic Review and Meta-Analysis of the Accuracy of in Vivo Reflectance Confocal Microscopy for the Diagnosis of Primary Basal Cell Carcinoma

Mihai Lupu [1,*], Iris Maria Popa [2], Vlad Mihai Voiculescu [1,3,*], Ana Caruntu [4,5] and Constantin Caruntu [6,7]

1. Department of Dermatology, "Carol Davila" University of Medicine and Pharmacy, 050474 Bucharest, Romania
2. Department of Plastic and Reconstructive Surgery, "Bagdasar-Arseni" Clinical Emergency Hospital, 041915 Bucharest, Romania; irismpopa@gmail.com
3. Department of Dermatology, "Elias" University Emergency Hospital, 011461 Bucharest, Romania
4. Department of Oral and Maxillofacial Surgery, "Carol Davila" Central Military Emergency Hospital, 010825 Bucharest, Romania; ana.caruntu@gmail.com
5. "Titu Maiorescu" University, Faculty of Medicine, 031593 Bucharest, Romania
6. Department of Dermatology, "Prof. N. Paulescu" National Institute of Diabetes, Nutrition and Metabolic Diseases, 011233 Bucharest, Romania; costin.caruntu@gmail.com
7. Department of Physiology, "Carol Davila" University of Medicine and Pharmacy, 050474 Bucharest, Romania
* Correspondence: lupu.g.mihai@gmail.com (M.L.); voiculescuvlad@yahoo.com (V.M.V.);
  Tel.: +40-74-0237-450 (M.L.); +40-72-2740-438 (V.M.V.)

Received: 3 September 2019; Accepted: 12 September 2019; Published: 13 September 2019

Abstract: Basal cell carcinoma (BCC) is the most common cancer worldwide and its incidence is constantly rising. Early diagnosis and treatment can significantly reduce patient morbidity and healthcare costs. The value of reflectance confocal microscopy (RCM) in non-melanoma skin cancer diagnosis is still under debate. This systematic review and meta-analysis were conducted to assess the diagnostic accuracy of RCM in primary BCC. PubMed, Google Scholar, Scopus, and Web of Science databases were searched up to July 05, 2019, to collect articles concerning primary BCC diagnosis through RCM. The studies' methodological quality was assessed by the QUADAS-2 tool. The meta-analysis was conducted using Stata 13.0, RevMan 5.0, and MetaDisc 1.4 software. We included 15 studies totaling a number of 4163 lesions. The pooled sensitivity and specificity were 0.92 (95% CI, 0.87–0.95; $I^2 = 85.27\%$) and 0.93 (95% CI, 0.85–0.97; $I^2 = 94.61\%$), the pooled positive and negative likelihood ratios were 13.51 (95% CI, 5.8–31.37; $I^2 = 91.01\%$) and 0.08 (95% CI, 0.05–0.14; $I^2 = 84.83\%$), and the pooled diagnostic odds ratio was 160.31 (95% CI, 64.73–397.02; $I^2 = 71\%$). Despite the heterogeneity and risk of bias, this study demonstrates that RCM, through its high sensitivity and specificity, may have a significant clinical impact on the diagnosis of primary BCC.

Keywords: basal cell carcinoma; diagnostic test accuracy; in vivo; reflectance confocal microscopy; systematic review; meta-analysis

1. Introduction

A significant increase in the worldwide incidence and prevalence of skin cancer, and especially basal cell carcinoma (BCC), has been reported in recent years [1–4]. Although locally invasive, this keratinocyte carcinoma has an excellent prognosis when diagnosed and treated early.

The routine diagnosis of BCC is based on clinical evaluation and histopathological examination, however with several caveats to this practice. Clinical diagnosis relies on the experience of the dermatologist and is subject to observer bias, and histopathological examination requires an invasive procedure prone to unavoidable sampling errors [5], sometimes requiring several interventions until a final diagnosis is reached.

Multiple techniques that enable non-invasive, real-time diagnosis of skin tumors have been developed, including dermoscopy, high-frequency ultrasonography [6], optical coherence tomography, multi-modal imaging [7], and reflectance confocal microscopy (RCM) [8–10]. RCM enables in vivo, non-invasive imaging of the skin layers and cellular structures in a horizontal plane at quasi-histologic resolution [11]. This imaging technique has been widely used in the diagnosis [12–20] and therapeutic monitoring [21–25] of skin cancer and inflammatory [26–30] and infectious skin diseases [31–33]. Numerous studies have investigated the diagnostic accuracy of in vivo RCM for BCC.

To formulate comprehensive and up-to-date evidence-based suggestions for the rational use of RCM, we performed a systematic review and meta-analysis to evaluate its accuracy in the diagnosis of primary BCC using histopathology as the reference standard.

2. Materials and Methods

A systematic review and meta-analysis was conducted and the results were reported according to the Preferred Reporting Items for Systematic Reviews and Meta-Analysis (PRISMA) statement [34]. Adjustments were made as to adhere to the recommendations for reviewing diagnostic test accuracy reports [35]. Because this study did not directly involve patients, an ethics committee approval was not required.

2.1. Study Objective and Definition of Reference Standard

The main objective of this systematic review and meta-analysis is to evaluate the accuracy of in vivo RCM for the diagnosis of primary BCC. A BCC diagnosis following histopathological examination of an incisional or excisional biopsy specimen was considered the reference standard.

2.2. Literature Search Strategy

One reviewer (ML) searched the following databases from inception till 05.07.2019: PubMed (keywords "(basal cell carcinoma) AND confocal microscopy"), Google Scholar (keywords "basal cell carcinoma" AND "confocal microscopy" -"ex vivo" -"ex-vivo", patents excluded), Web of Science (keywords "TS = (confocal microscopy AND basal cell carcinoma)Timespan: All years. Indexes: SCI-EXPANDED, SSCI, A&HCI, CPCI-S, CPCI-SSH, BKCI-S, BKCI-SSH, ESCI, CCR-EXPANDED, IC.") and Elsevier SCOPUS (keywords "TITLE-ABS-KEY ("confocal microscopy" AND "basal cell carcinoma")"). All references were imported and deduplicated using the reference manager EndNote (version X7, 1988–2013 Thomson-Reuters). Only articles written in English were taken into account for inclusion.

2.3. Eligibility Criteria

Two reviewers (ML and VMV) screened all retrieved articles by title and abstract to establish their relevance. Full-text recovery and analysis were done only for potentially eligible articles. Disagreements were settled through discussion with a third reviewer (MIP).

The established eligibility criteria were: (1) the RCM device used in the study was the VivaScope 1000 or 1500 (Lucid Technologies, Henrietta, NY, USA; Caliber I.D., Rochester, NY, USA); (2) the investigated lesions were primary BCCs, any histopathological subtype; (3) the reference standard was a diagnosis of BCC following the histopathological examination of incisional or excisional biopsy specimen; (4) sufficient data for the reconstruction of a 2×2 table or specified values for sensitivity (Sn) and specificity (Sp) were available.

We excluded from the analysis: (1) reviews, editorials, opinions, ex-vivo studies; (2) clinical cases or case series including less than 10 BCCs, in order to avoid a small studies effect; (3) studies were full-text and recovery was not possible, even after searching the available medical databases and/or contacting the corresponding authors. Studies thought to include overlapping populations were also excluded, keeping only the one with the largest number of participants. Additionally, the reference list of each study was checked to identify further relevant articles that may have been overlooked during initial screening.

2.4. Data Extraction and Quality Evaluation of the Studies

One reviewer (ML) extracted the data from the included studies into a predefined form, validated by another reviewer (CC). The following parameters were extracted: the name of the first author, year of publication, country, number of participating centers, study type (prospective/retrospective), lesion type, number of investigators and their experience level (low/high), RCM device, total number of patients and lesions, patient gender and age (mean/median, years), confocal criteria employed for the diagnosis of BCC, number of true and false positives and negatives.

All included articles were evaluated using the QUADAS-2 (Quality Assessment of Diagnostic Accuracy Studies) tool, which has a maximum score of 14 points [36]. QUADAS-2 offers a perspective over the methodological quality of a study through the assessment of four key domains: patient selection, index test (in vivo RCM), reference standard (histopathological examination), and patient flow and timing in the study. Each of these domains is evaluated for risk of bias, while the first three domains are also evaluated regarding applicability concerns.

2.5. Statistical Analysis and Meta-Analysis

Two-by-two tables were constructed for each RCM-based diagnosis of BCC against histopathology from incisional or excisional biopsy specimens and sensitivity, specificity and their 95% confidence intervals were visually represented using forest plots.

We used a bivariate model (hierarchical logistic regression) for the meta-analysis of sensitivity and specificity and to create the HSROC (summary receiver operating characteristic) curve. The HSROC curve illustrates sensitivity versus specificity and supplies information regarding the overall test performance across different thresholds. This model accounted for both the within- and between-study variability.

Every meta-analysis of diagnostic accuracy tests suffers from heterogeneity, attributed mainly to index test efficiency variation due to different diagnostic thresholds. Therefore, we considered the investigation of heterogeneity sources outweighs the mere demonstration of its existence [37]. Heterogeneity sources were evaluated through subgroup analyses and meta-regression using the following variables: study type (prospective/retrospective), reference standard (incisional/excisional biopsy), RCM device (VivaScope 1000/1500) and investigator experience level (low/high). Deeks asymmetry test and funnel plot were used to evaluate publication bias [38].

Data organization and statistical analyses were carried out using the software packages STATA (v13.0; StataCorp LP, Texas, USA), MetaDisc (v1.4; Unidad de Bioestadistica Clinica—Hospital Ramon y Cajal, Universidad Complutense, Madrid, Spain) and Review Manager (v5.3; Nordic Cochrane Center, Copenhagen, Denmark).

3. Results

3.1. Literature Search Results

The initial database search identified a total number of 4624 items. After deduplication, only 3627 remained. After title and abstract evaluation 3543 items were excluded and only 84 were selected for full-text retrieval and analysis. Sixty-nine articles were excluded based on full-text analysis (motives were recorded) (Figure 1). Fifteen studies totaling a number of 4163 lesions were included in the final analysis [5,19,39–51]. Study characteristics were summarized in Table 1.

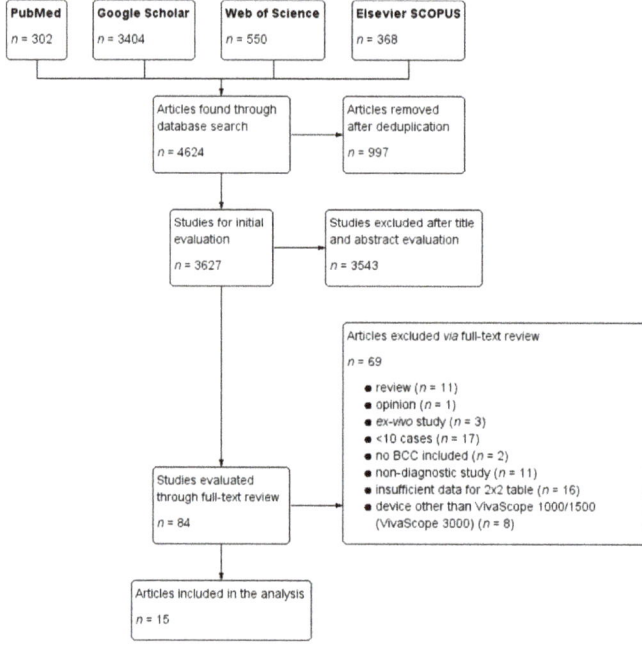

Figure 1. Screening process and results. Basal cell carcinoma (BCC).

The male/female ratio could not be calculated due to missing data in several studies. The manufacturer of the RCM devices VivaScope 1000 and 1500 was Lucid Inc. (Lucid Technologies, Henrietta, NY, USA), the majority of studies being carried out in Europe. A single study [39] utilized a prototype version of the VivaScope 1000 (Wellman Laboratories, Boston, MA, USA) and in two multicenter studies [41,42] different RCM devices were used, according to each participating center. Three studies did not specify the investigators' level of experience with RCM [43,45,50]. Confocal criteria for BCC diagnosis varied considerably between studies (Table 2).

Table 1. Characteristics of included studies.

Author, Year, [Reference]	Country	No. of Centers	Study Design	Types of Lesion	No. of Investigators	Experience Level	Reference Standard	RCM device	No. of Patients (M/F)	Age (Mean/Median)	No. of Lesions
Castro et al. 2015 [46]	Brazil&USA	2	prospective	BCC	2	low	histopathology (incisional)	VivaScope 1500	32 (20/12)	65	54
Gerger et al. 2006 [40]	Austria	1	prospective	melanoma, BCC, nevi, SebK	4	low	clinic & histopathology (excisional)	VivaScope 1500	119 (62/57)	n/a	120
Guitera et al. 2012 [41]	Australia&Italy	2	prospective	melanoma, BCC, SCC, nevi	2	high	histopathology (excisional)	VivaScope 1000&1500	663 (354/309)	53	710
Longo et al. 2013 [42]	Italy	2	retrospective	melanoma, BCC, SCC, nevi, SebK, DF	1	high	histopathology (n/a)	VivaScope 1000&1500	140 (64/76)	50	140
Nori et al. 2004 [39]	USA&Spain	4	retrospective	BCC, various others	1	low	clinical & histopathology (incisional)	VivaScope 1000 & Wellman Laboratories prototype	145 (n/a)	n/a	152
Peppelman et al. 2013 [43]	Netherlands	1	prospective	BCC	n/a	n/a	histopathology (incisional)	VivaScope 1500	27 (16/11)	66	57
Rao et al. 2013 [44]	USA	1	retrospective	melanoma, BCC, various benign	2	low	histopathology (incisional)	VivaScope 1500	n/a	n/a	334
Pellacani et al. 2014 [45]	Italy	1	prospective	melanoma, BCC, various benign	1	n/a	histopathology (excisional)	VivaScope 1500	408	41	292
Farnetani et al. 2015 [47]	Italy	1	retrospective	melanoma, BCC, AKs, various benign	9	high & low	histopathology (n/a)	VivaScope 1500	n/a	n/a	100
Guitera et al. 2016 [48]	Australia&Italy	3	retrospective	melanoma, BCC, AKs, various benign	1	high	histopathology (excisional)	VivaScope 1500	n/a	54.8	191

Table 1. Cont.

Author, Year, [Reference]	Country	No. of Centers	Study Design	Types of Lesion	No. of Investigators	Experience Level	Reference Standard	RCM device	No. of Patients (M/F)	Age (Mean/Median)	No. of Lesions
Kadouch et al. 2017 [5]	Netherlands	2	prospective	BCC	2	low	histopathology (excisional)	VivaScope 1500	46	64	46
Nelson et al. 2016 [49]	USA	1	prospective	BCC	8	low	histopathology (biopsy)	VivaScope 1500	87 (65/22)	73	100
Witkowski et al. 2015 [50]	Italy	1	retrospective	BCC, melanoma, SCC, various benign	1	n/a	histopathology (n/a)	VivaScope 1500	n/a	n/a	260
Peccerillo et al. 2018 [51]	Italy	1	retrospective	BCC, melanoma, SCC, SebK, DF	2	high	histopathology (n/a)	VivaScope 1500	n/a	n/a	1484
Lupu et al. 2019 [19]	Romania	2	retrospective	BCC, SCC, AKs, Bowen's disease, various benign	2	high	histopathology (excisional)	VivaScope 1500	87 (36/51)	68.1	123

BCC, basal cell carcinoma; SCC, squamous cell carcinoma; SebK, seborrheic keratoses; AKs, actinic keratoses; DF, dermatofibroma; n/a, not available.

Table 2. Criteria for the diagnosis of basal cell carcinoma in the included studies.

Author, Year, [Reference]	Reflectance Confocal Microscopic Criteria
Castro et al. 2015 [46]	hyporefractile silhouettes, tumor islands, epidermal streaming, peripheral palisading, peri-tumoral clefting, peri-tumoral collagen bundles, increased vascularization, dendritic structures
Gerger et al. 2006 [40]	increased vascularization, epidermal streaming, peri-tumoral collagen bundles
Guitera et al. 2012 [41]	epidermal streaming, dilated horizontal blood vessels, basaloid cord or nodule, epidermal shadow, glomerular vessels, non-visible dermal papillae, epidermal disarray, dendritic structures, peri-tumoral clefting, cells with visible nuclei inside tumor islands
Longo et al. 2013 [42]	epidermal disarray, ulceration or erosion, cauliflower architecture, hyporefractile silhouettes, bright filaments inside tumor islands, increased vascularization, inflammatory infiltrate
Nori et al. 2004 [39]	elongated monomorphic nuclei, inflammatory infiltrate, increased vascularization, epidermal pleomorphism
Peppelman et al. 2013 [43]	tumor islands, peri-tumoral clefting, peripheral palisading, elongated and polarized nuclei, keratinocyte atypia and spongiosis, solar elastosis, increased vascularization, inflammatory infiltrate, leukocyte rolling
Rao et al. 2013 [44]	diagnostic criteria not specified
Pellacani et al. 2014 [45]	diagnostic criteria not specified
Farnetani et al. 2015 [47]	basaloid cords, ulceration, disarray at the dermal-epidermal junction
Guitera et al. 2016 [48]	epidermal streaming, basaloid cord or nodule, peri-tumoral fibrillar polarized pattern, peri-tumoral clefting, epidermal shadow, dark nodules, dilated horizontal blood vessels, glomerular vessels
Kadouch et al. 2017 [5]	diagnostic criteria not specified
Nelson et al. 2016 [49]	tumor islands, peri-tumoral clefting, hyporefractile silhouettes, canalicular vessels, dendritic cells
Witkowski et al. 2015	diagnostic criteria not specified
Peccerillo et al. 2018 [51]	mild keratinocyte atypia, epidermal streaming, streaming epidermis, cords connected to the epidermis, dark silhouettes, peri-tumoral clefts, ulceration/erosion, tumor island size and location (epidermal or dermal), branch-like structures in tumor island, peripheral palisading, vascular morphology (linear or coiled vessels) and diameter, collagen surrounding tumor islands, solar elastosis and inflammatory infiltrates
Lupu et al. 2019 [19]	keratinocyte atypia, epidermal streaming, ulceration, cords connected to the epidermis, small tumor islands (diameter <300 m), large tumor islands (diameter >300 m), hyporefractile silhouettes, peripheral palisading, clefting, increased vascularization, "onion-like" structures, peri-tumoral collagen bundles, inflammation represented by bright dots and plump bright cells, and dendritic cells inside tumor islands

3.2. Quality Assessment of Study Reports

The results of the methodological quality assessment of the studies are illustrated in Figures 2 and 3.

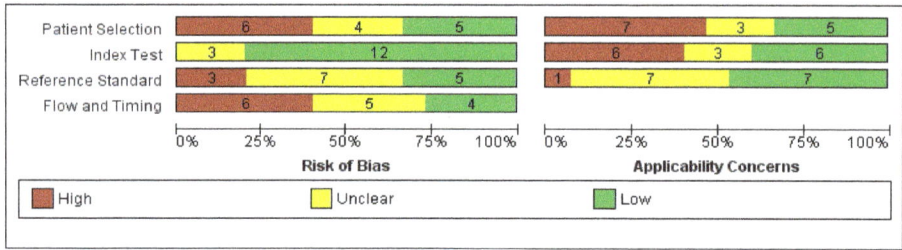

Figure 2. Included studies according to QUADAS-2 guidelines.

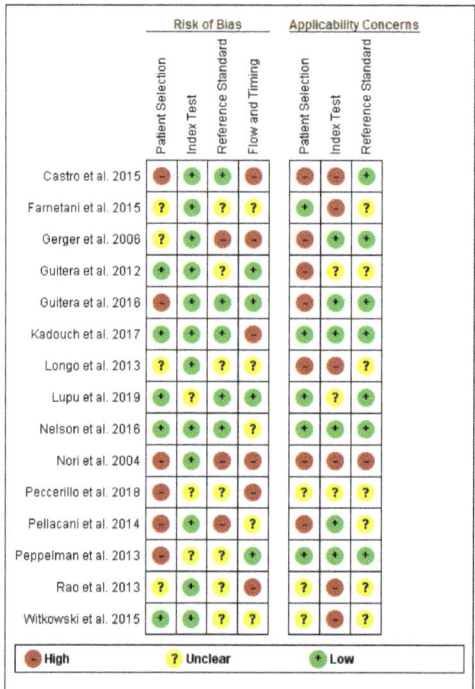

Figure 3. Methodological quality assessment via QUADAS-2 tool.

Eight studies had a retrospective design, while only seven were prospective. In general, the included studies exhibited high or unclear risk for bias in all domains except the index test and high or unclear applicability concerns. Ten studies (66.66%) had a high (n = 6) or unclear (n = 4) risk of bias concerning patient selection, mostly due to the exclusion of poor quality images, case-control design or unspecified patient selection protocol. Only five studies fully described the patient selection protocol. Ten studies presented high (n = 7) or uncertain (n = 3) applicability concerns owing to restrictions applied to the studied population (only including lesions highly suspicious of BCC, only including nodular lesions, etc.) and inclusion of patients with multiple lesions. In their retrospective study, Longo et al. [42] only included histopathologically confirmed nodular lesions, compensating through a relatively large sample (n = 140) and a wide variety of lesions. Peccerillo et al. [51] only included dermoscopically equivocal pigmentary lesions and excluded lesions located on the face, again compensating through a very large sample size (n = 1484). Castro et al. [46], Longo et al. [42], and Peppelman et al. [43] excluded lesions which, based on their location or the presence of hyperkeratosis, could not be evaluated by RCM and lesions in which RCM evaluation was inconclusive. Although understandable why lesions not suitable for RCM examinations due to physical limitations may not be included, these exclusions could have led to an overestimation of specificity.

Twelve out of the 15 included studies had a low risk of bias concerning the index test. More than half (n = 9) of the studies had high or uncertain applicability concerns in the index test domain due to tele-diagnosis use, blinding of the investigators to patient history or clinical data, presentation only of diagnostic consensus or lack of a diagnostic threshold.

Five studies had a low risk of bias regarding the use of the reference standard, while three were at high risk of bias owing to inadequate reference standards. Seven studies were at an unclear risk of bias. In two studies [39,40], not all lesions underwent histopathological examination. Regarding applicability concerns of the reference standard, only one study [39] had a high risk owing to the use of expert clinical diagnosis as a reference standard, while seven studies did not specify the pathologists' experience level. Although the excision of all benign lesions included in a study is not practical, studies in which a clinical diagnosis was designated as definitive were considered as having a high risk of bias.

Regarding flow and timing according to the QUADAS-2 tool, six studies had a high risk of bias, while five and four studies had unclear and low risk of bias, respectively. Gerger et al. [40], Guitera et al. [41], Lupu et al. [19], Peccerillo et al. [51], and Longo et al. [42] included patients suspected of skin cancer (including melanoma) which could have simplified the diagnosis of basal cell carcinoma, however all studies included a fair number of both benign and malignant lesions somewhat compensating for this limitation. Nori et al. [39], Gerger et al. [40], Rao et al. [44], Peccerillo et al. [51], and Castro et al. [46] did not specify the time interval between index test (RCM) and reference standard (histopathological examination).

3.3. Diagnostic Accuracy of RCM and Meta-Analysis

All fifteen studies were included in the meta-analysis. Sensitivity ranged from 73% to 100%, while specificity ranged from 38% to 100%. The pooled sensitivity and specificity values were 0.92 (95% CI, 0.87–0.95; I^2 = 85.27%) and 0.93 (95% CI, 0.85–0.97; I^2 = 94.61%). The distributions of RCM sensitivity and specificity and their summary values for the diagnosis of BCC in the included studies is represented in Figure 4.

The positive likelihood ratio ranged from 1.62 (95% CI, 0.96–2.72) to 2315.51 (95% CI, 144.33–37148.9) and the negative likelihood ratio ranged from 0.011 (95% CI, 0.001–0.17) to 0.3 (95% CI, 0.19–0.49). The pooled positive and negative likelihood ratios were 13.51 (95% CI, 5.8–31.37; I^2 = 91.01%) and 0.08 (95% CI, 0.05–0.14; I^2 = 84.83%). The diagnostic odds ratio (DOR) ranged from 21.37 (95% CI, 9.39–48.61) to 12725 (95% CI, 508.97–318141.1). The pooled DOR was 160.31 (95% CI, 64.73–397.02; I^2 = 71%).

The shape of the HSROC curve in Figure 5 and the area under the curve (AUC) of 0.97 suggested the lack of a threshold effect. The shape of the prediction region is meant to give a graphic representation of the extent of between-study heterogeneity, is dependent on the assumption of a bivariate normal distribution for the random effects, and should therefore not be over-interpreted [52].

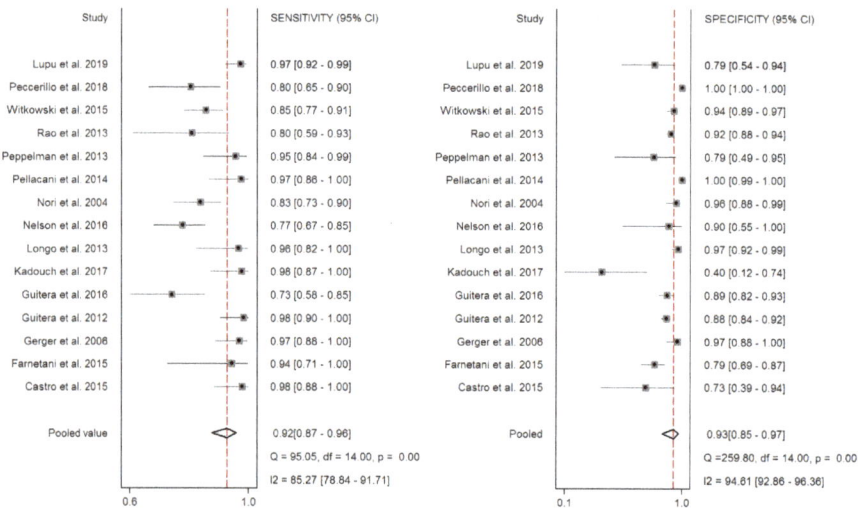

Figure 4. Forest plots for individual studies and pooled estimates of sensitivity and specificity with corresponding heterogeneity statistics of reflectance confocal microscopy for the diagnosis of basal cell carcinoma.

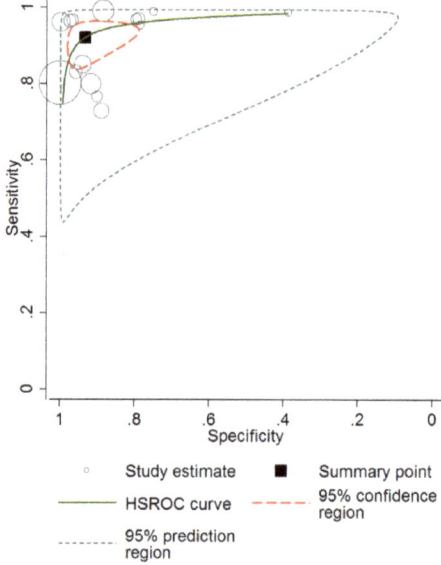

Figure 5. Curve summarizing reflectance confocal microscopy (RCM) sensitivity and specificity for BCC diagnosis.

3.4. Heterogeneity Analysis

Concerning heterogeneity analysis, a Spearman correlation coefficient of 0.468 ($p = 0.079$) suggested the lack of a threshold effect.

Next, we investigated potential sources of heterogeneity, other than the threshold effect. We performed a meta-regression analysis employing the following covariates as predictors: (1) study design (prospective/retrospective), (2) RCM device (VivaScope 1000/1500), (3) reference standard (histopathology from incisional/excisional biopsy specimen), (4) investigator experience level (low/high), and (5) number of participating centers (single center/multicenter).

The results showed that a prospective study design was associated with a 9.35 times higher RCM diagnostic performance compared with the retrospective design (RDOR = 9.35; 95% CI, 1.17;74.56; p = 0.037), while using the histopathology examination of the excisional biopsy specimen as a reference standard resulted in a 3.27 times (RDOR = 3.27; 95% CI, 0.93;11.47; p = 0.06) higher index test performance. The type of RCM device, investigator experience, and number of participating centers were not significant predictors in our meta-regression model (p = 0.46, 0.91 and 0.5, respectively). The results of the meta-regression are summarized in Table 3.

Table 3. Results of the meta-regression for heterogeneity sources.

Covariate	Coefficient	Standard Error	p	RDOR	(95% CI)
Study design	2.236	0.9	0.037	9.35	(1.17; 74.56)
RCM device	−0.838	1.09	0.46	0.43	(0,03; 5.38)
Reference standard	1.184	0.54	0.06	3.27	(0.93; 11.47)
Investigator experience	0.067	0.59	0.91	1.07	(0.27; 4.2)
Number of centers	0.561	0.79	0.5	1.75	(0.28; 10.98)

RDOR, Relative Diagnostic Odds Ratio; RCM, reflectance confocal microscopy.

Subgroup analysis revealed that RCM pooled sensitivity and specificity values in the retrospective study designs (n = 8) were 0.87 (95% CI, 0.796–0.926) and 0.95 (95% CI, 0.855–0.983) compared to 0.95 (95% CI, 0.895–0.982) and 0.90 (95% CI, 0.689–0.974) in the prospective study designs (n = 7). The pooled positive and negative likelihood ratios in retrospective studies were 17.55 (95% CI, 5.91–52.06) and 0.131 (95% CI, 0.08–0.215). The same ratios were 9.67 (95% CI, 2.73–34.27) and 0.048 (95% CI, 0.02–0.115) in prospective studies. The graphical representation of the diagnostic odds ratios (DOR) along with standard errors and confidence intervals for each study are illustrated in Figure 6.

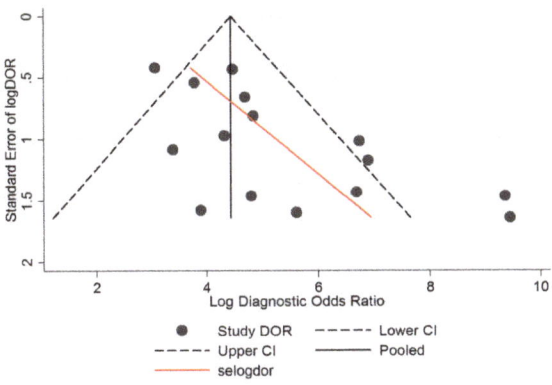

Figure 6. Plot with pseudo 95% confidence limits in the included studies.

Finally, we sought to identify potential publication bias. The funnel plot of Deeks asymmetry test [38] was relatively symmetrical (Figure 7), suggesting the lack of publication bias (p = 0.45).

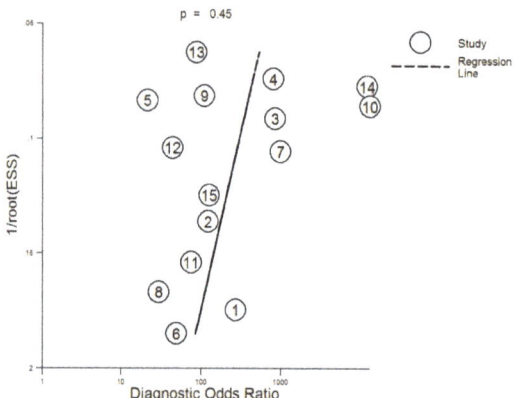

Figure 7. Plot of Deeks asymmetry test for publication bias.

Although we chose to report the results of the meta-analysis, they should be interpreted exercising caution and keeping in mind its limitations due to variation and potential biases.

4. Discussion

RCM is a novel, non-invasive diagnostic technique that enables real-time imaging of the skin down to the upper layers of the dermis at resolutions similar to histology. The confocal criteria for RCM diagnosis of various skin tumors are relatively easy to learn and the results are reproducible [53].

This systematic review and meta-analysis compares the diagnostic accuracy of RCM to histopathological examination from an incisional or excisional biopsy specimen using the results of 15 studies which included a total number of 4163 lesions. Our literature search strategy used broad keywords in multiple databases to identify as many studies as possible.

The results of the meta-analysis show a sensitivity of 92% and a specificity of 93% for the in vivo RCM diagnosis of BCC. However, these high values of both sensitivity and specificity must be interpreted with caution. The significant amount of heterogeneity renders the direct comparison of RCM diagnostic accuracy between studies impossible. RCM sensitivity for the diagnosis of BCC ranged between 73% and 100%, and its specificity ranged between 38% and 100%. Although statistically non-significant (possibly due to insufficient statistical power), these wide variations could still be attributed to the different confocal criteria and slightly different reference standards (incisional versus excisional biopsy specimen), but also investigator experience, and possibly other unknown heterogeneity sources. Investigator experience could influence diagnostic accuracy even when using the same diagnostic criteria. Rao et al. demonstrated a higher sensitivity (97.4% vs. 93.1%) and specificity (80.5% vs. 64.1%) for an investigator with over nine years of experience with RCM compared to one with only one year experience [44].

We observed that the RCM performance in prospective studies was significantly superior to that of retrospective studies (prospective vs. retrospective, RDOR = 9.35, $p = 0.037$). The pooled specificities of prospective and retrospective studies were consistent (90% vs. 95%), but the sensitivity for prospective studies was higher than that for retrospective ones (95.6% vs. 87.52%). Although the results of prospective studies were more reliable, a variety of uncontrollable factors, such as RCM devices and software and investigator experience may still influence the final diagnostic accuracy.

Subgroup analysis revealed that RCM pooled sensitivity and specificity values in the retrospective study designs (n = 8) were 0.87 (95% CI, 0.796–0.926) and 0.95 (95% CI, 0.855–0.983) compared to 0.95 (95% CI, 0.895–0.982) and 0.90 (95% CI, 0.689–0.974) in the prospective study designs (n = 7).

5. Clinical Relevance

The results of this study may have significant implications for patients suffering from BCC. Based on recent epidemiological data, the expected prevalence of a primary BCC in Europe is 1.4% [54,55]. Using this available data together with our results, the absolute number of true and false positives and negatives can be estimated in a hypothetical cohort of 1000 subjects. This means that 14 subjects in this cohort would have a primary BCC. By using RCM as a diagnostic tool with a sensitivity of 92% and a specificity of 93%, just one of these 14 BCCs would go unnoticed, while 69 patients would be unnecessarily treated (Figure 8).

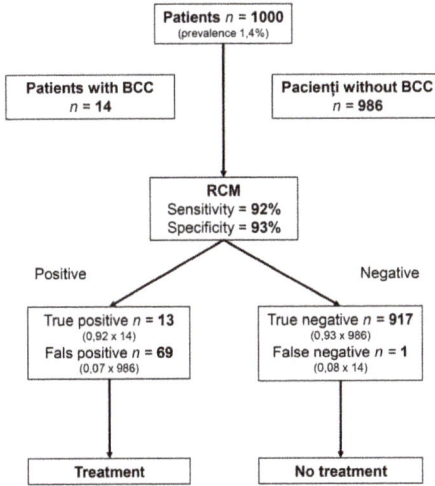

Figure 8. The consequences of using reflectance confocal microscopy for BCC diagnosis in a cohort of 1000 subjects. The use of RCM would lead to 82 patients being treated, of which 69 would not need to be treated; 918 patients would not be treated, of which only one would have necessitated treatment. *BCC, basal cell carcinoma; RCM, reflectance confocal microscopy.*

In vivo RCM could therefore become a very useful technique in the diagnosis of BCC. However, in order for it to be regarded as a potential replacement for histopathological examination, this non-invasive technique should have the ability to discriminate between the different histopathological BCC subtypes [56]. This aspect is of critical importance due to the different therapeutic approaches to BCC based on its histopathological subtype [57]. Several studies, some of which are included in this analysis [5,19,43] have sought to determine specific RCM criteria for the discrimination of BCC histotype. Unfortunately, we were unable to estimate sensitivity and specificity of BCC subtyping through in vivo RCM from the data available in the included studies.

6. Strengths and Limitations

We consider the adherence to the PRISMA guidelines [34], the rigorous examination of the existing literature, and the use of the QUADAS-2 tool [36] for methodological quality assessment to be strengths of our analysis.

Our results should be interpreted bearing in mind some limitations: the relatively small number of studies (n = 15) included in the analysis; the double reference standard (histopathological examination from incisional and excisional biopsy specimen; ideally, only the excisional biopsy specimen should be used); the incomplete reporting of the patient selection process in some studies; the use of different confocal criteria for the diagnosis of BCC; the variation in RCM device and investigator experience between studies. Regarding the confocal criteria for BCC diagnosis, an international consensus for use in future studies is desirable. To facilitate homogeneity, futurestudies could consider reporting investigator experience in years, number of examined lesions and/or attended courses.

7. Future Directions

We expect more studies investigating the diagnostic accuracy of in vivoreflectance confocal microscopy for BCC will be carried out. To promote comparability of their results, future studies should adhere to STARD guidelines [58] and use the histopathological examination of the excisional biopsy specimen as a reference standard.

Moreover, as this non-invasive technique becomes more widely disseminated, studies could benefit from the use of RCM devices with similar technical properties and standardization of imaging protocols. To assure results comparability, these studies should report the investigators' level of experience with RCM. More studies that investigate RCM accuracy for BCC histopathological subtype are needed. Additionally, comparative studies analyzing the cost/efficiency ratio between RCM and the current standard (histopathological examination of the incisional biopsy specimen) are warranted.

8. Conclusions

Reflectance confocal microscopy is a promising technique in the diagnosis of primary basal cell carcinoma. A definitive conclusion could only be drawn when a higher number of studies, possibly with homogeneous methodological approach, will be available.

Author Contributions: M.L., C.C. and V.M.V. contributed to the conception of this study and performed the preliminary documentation. All authors participated in the design of the study and implemented the research. M.L., C.C., I.M.P. and V.M.V. were responsible for the data acquisition, selection and analysis, and clinical interpretation of the data. M.L., I.M.P., V.M.V., A.C. and C.C. participated in the statistical analysis and contributed to the interpretation of the results as well as the manuscript drafting and writing of the study. M.L., V.M.V. and C.C. have revised critically the manuscript for important intellectual content. All authors reviewed and approved the final manuscript.

Funding: This research and APC were funded by a grant of Romanian Ministry of Research and Innovation, CCCDI-UEFISCDI, [project number 61PCCDI/2018 PN-III-P1-1.2-PCCDI-2017-0341], within PNCDI-III.

Conflicts of Interest: The authors declare no conflict of interest. The funders had no role in the design of the study; in the collection, analyses, or interpretation of data; in the writing of the manuscript, or in the decision to publish the results.

References

1. Birch-Johansen, F.; Jensen, A.; Mortensen, L.; Olesen, A.B.; Kjaer, S.K. Trends in the incidence of nonmelanoma skin cancer in denmark 1978–2007: Rapid incidence increase among young danish women. *Int. J. Cancer* **2010**, *127*, 2190–2198. [CrossRef] [PubMed]
2. Miller, D.L.; Weinstock, M.A. Nonmelanoma skin cancer in the united states: Incidence. *J. Am. Acad. Dermatol.* **1994**, *30*, 774–778. [CrossRef]
3. Trakatelli, M.; Ulrich, C.; del Marmol, V.; Euvrard, S.; Stockfleth, E.; Abeni, D. Epidemiology of nonmelanoma skin cancer (nmsc) in europe: Accurate and comparable data are needed for effective public health monitoring and interventions. *Br. J. Dermatol.* **2007**, *156* (Suppl. 3), 1–7. [CrossRef] [PubMed]
4. Papagheorghe, L.M.L.; Lupu, M.; Pehoiu, A.G.; Voiculescu, V.M.; Giurcaneanu, C. Basal cell carcinoma—Increasing incidence leads to global health burden. *RoJCED* **2015**, *2*, 106–111.
5. Kadouch, D.J.; Leeflang, M.M.; Elshot, Y.S.; Longo, C.; Ulrich, M.; van der Wal, A.C.; Wolkerstorfer, A.; Bekkenk, M.W.; de Rie, M.A. Diagnostic accuracy of confocal microscopy imaging vs. Punch biopsy for diagnosing and subtyping basal cell carcinoma. *J. Eur. Acad. Dermatol. Venereol.* **2017**, *31*, 1641–1648. [CrossRef] [PubMed]
6. Kučinskienė, V.; Samulėnienė, D.; Gineikienė, A.; Raišutis, R.; Kažys, R.; Valiukevičienė, S. Preoperative assessment of skin tumor thickness and structure using 14-mhz ultrasound. *Medicina* **2014**, *50*, 150–155. [CrossRef] [PubMed]

7. Heuke, S.; Vogler, N.; Meyer, T.; Akimov, D.; Kluschke, F.; Röwert-Huber, H.-J.; Lademann, J.; Dietzek, B.; Popp, J. Detection and discrimination of non-melanoma skin cancer by multimodal imaging. *Healthcare* **2013**, *1*, 64–83. [CrossRef] [PubMed]
8. Fahradyan, A.; Howell, A.C.; Wolfswinkel, E.M.; Tsuha, M.; Sheth, P.; Wong, A.K. Updates on the management of non-melanoma skin cancer (nmsc). *Healthcare* **2017**, *5*, 82. [CrossRef] [PubMed]
9. Diaconeasa, A.; Boda, D.; Neagu, M.; Constantin, C.; Căruntu, C.; Vlădău, L.; Guţu, D. The role of confocal microscopy in the dermato–oncology practice. *J. Med. Life* **2011**, *4*, 63.
10. Ilie, M.A.; Caruntu, C.; Lupu, M.; Lixandru, D.; Georgescu, S.-R.; Bastian, A.; Constantin, C.; Neagu, M.; Zurac, S.A.; Boda, D. Current and future applications of confocal laser scanning microscopy imaging in skin oncology. *Oncol. Lett.* **2019**. [CrossRef]
11. Rajadhyaksha, M.; Grossman, M.; Esterowitz, D.; Webb, R.H.; Anderson, R.R. In vivo confocal scanning laser microscopy of human skin: Melanin provides strong contrast. *J. Invest. Dermatol.* **1995**, *104*, 946–952. [CrossRef] [PubMed]
12. Caruntu, C.; Boda, D.; Gutu, D.E.; Caruntu, A. In vivo reflectance confocal microscopy of basal cell carcinoma with cystic degeneration. *Rom. J. Morphol. Embryol.* **2014**, *55*, 1437–1441. [PubMed]
13. Ghita, M.A.; Caruntu, C.; Rosca, A.E.; Kaleshi, H.; Caruntu, A.; Moraru, L.; Docea, A.O.; Zurac, S.; Boda, D.; Neagu, M.; et al. Reflectance confocal microscopy and dermoscopy for in vivo, non-invasive skin imaging of superficial basal cell carcinoma. *Oncol. Lett.* **2016**, *11*, 3019–3024. [CrossRef] [PubMed]
14. Vaišnorienė, I.; Rotomskis, R.; Kulvietis, V.; Eidukevičius, R.; Žalgevičienė, V.; Laurinavičienė, A.; Venius, J.; Didžiapetrienė, J. Nevomelanocytic atypia detection by in vivo reflectance confocal microscopy. *Medicina* **2014**, *50*, 209–215. [CrossRef] [PubMed]
15. Lupu, M.; Caruntu, C.; Solomon, I.; Popa, A.; Lisievici, C.; Draghici, C.; Papagheorghe, L.; Voiculescu, V.M.; Giurcaneanu, C. The use of in vivo reflectance confocal microscopy and dermoscopy in the preoperative determination of basal cell carcinoma histopathological subtypes. *DermatoVenerol.* **2017**, *62*, 7–13.
16. Lupu, M.; Caruntu, A.; Caruntu, C.; Boda, D.; Moraru, L.; Voiculescu, V.; Bastian, A. Non-invasive imaging of actinic cheilitis and squamous cell carcinoma of the lip. *Mol. Clin. Oncol.* **2018**, *8*, 640–646. [CrossRef] [PubMed]
17. Lupu, M.; Căruntu, A.; Moraru, L.; Voiculescu, V.M.; Boda, D.; Tănase, C.; Căruntu, C. Non-invasive imaging techniques for early diagnosis of radiation-induced squamous cell carcinoma of the lip. *Rom. J. Morphol. Embryol.* **2018**, *59*, 3.
18. Lupu, M.; Căruntu, C.; Vâjâitu, C.; Solomon, I.; Voiculescu, V.M.; Popa, M.I.; Drăghici, C.; Giurcăneanu, C. In vivo reflectance confocal microscopy of spoke-wheel structures in a pigmented basal cell carcinoma. Case report. *DermatoVenerol* **2019**, *64*, 11–16.
19. Lupu, M.; Popa, I.M.; Voiculescu, V.M.; Boda, D.; Caruntu, C.; Zurac, S.; Giurcaneanu, C. A retrospective study of the diagnostic accuracy of in vivo reflectance confocal microscopy for basal cell carcinoma diagnosis and subtyping. *J. Clin. Med.* **2019**, *8*, 449. [CrossRef]
20. González, S. Clinical applications of reflectance confocal microscopy in the management of cutaneous tumors. *Actas Dermo-Sifiliogr.* **2008**, *99*, 528–531. [CrossRef]
21. Torres, A.; Niemeyer, A.; Berkes, B.; Marra, D.; Schanbacher, C.; Gonzalez, S.; Owens, M.; Morgan, B. 5% imiquimod cream and reflectance-mode confocal microscopy as adjunct modalities to mohs micrographic surgery for treatment of basal cell carcinoma. *Dermatol. Surg.* **2004**, *30*, 1462–1469. [PubMed]
22. Marra, D.E.; Torres, A.; Schanbacher, C.F.; Gonzalez, S. Detection of residual basal cell carcinoma by in vivo confocal microscopy. *Dermatol. Surg.* **2005**, *31*, 538–541. [CrossRef] [PubMed]
23. Ahlgrimm-Siess, V.; Horn, M.; Koller, S.; Ludwig, R.; Gerger, A.; Hofmann-Wellenhof, R. Monitoring efficacy of cryotherapy for superficial basal cell carcinomas with in vivo reflectance confocal microscopy: A preliminary study. *J. Dermatol. Sci.* **2009**, *53*, 60–64. [CrossRef] [PubMed]
24. Venturini, M.; Zanca, A.; Calzavara-Pinton, P. In vivo non-invasive evaluation of actinic keratoses response to methyl-aminolevulinate-photodynamic therapy (mal-pdt) by reflectance confocal microscopy. *Cosmetics* **2014**, *1*, 37–43. [CrossRef]
25. Gamble, R.G.; Jensen, D.; Suarez, A.L.; Hanson, A.H.; McLaughlin, L.; Duke, J.; Dellavalle, R.P. Outpatient follow-up and secondary prevention for melanoma patients. *Cancers* **2010**, *2*, 1178–1197. [CrossRef]

26. Ianoși, S.L.; Forsea, A.M.; Lupu, M.; Ilie, M.A.; Zurac, S.; Boda, D.; Ianosi, G.; Neagoe, D.; Tutunaru, C.; Popa, C.M. Role of modern imaging techniques for the in vivo diagnosis of lichen planus. *Exp. Ther. Med.* **2019**, *17*, 1052–1060. [CrossRef] [PubMed]
27. Caruntu, C.; Boda, D. Evaluation through *in vivo* reflectance confocal microscopy of the cutaneous neurogenic inflammatory reaction induced by capsaicin in human subjects. *J. Biomed. Opt.* **2012**, *17*, 085003. [CrossRef] [PubMed]
28. Căruntu, C.; Boda, D.; Căruntu, A.; Rotaru, M.; Baderca, F.; Zurac, S. In vivo imaging techniques for psoriatic lesions. *Rom. J. Morphol. Embryol.* **2014**, *55*, 1191–1196. [PubMed]
29. Lacarrubba, F.; Verzi, A.E.; Errichetti, E.; Stinco, G.; Micali, G. Darier disease: Dermoscopy, confocal microscopy, and histologic correlations. *J. Am. Acad. Dermatol.* **2015**, *73*, e97–e99. [CrossRef] [PubMed]
30. Ionescu, A.-M.; Ilie, M.-A.; Chitu, V.; Razvan, A.; Lixandru, D.; Tanase, C.; Boda, D.; Caruntu, C.; Zurac, S. In vivo diagnosis of primary cutaneous amyloidosis—The role of reflectance confocal microscopy. *Diagnostics* **2019**, *9*, 66. [CrossRef] [PubMed]
31. Nazzaro, G.; Farnetani, F.; Moltrasio, C.; Passoni, E.; Pellacani, G.; Berti, E. Image gallery: Demodex folliculorum longitudinal appearance with reflectance confocal microscopy. *Br. J. Dermatol.* **2018**, *179*, e230. [CrossRef] [PubMed]
32. Lacarrubba, F.; Verzì, A.E.; Micali, G. Detailed analysis of in vivo reflectance confocal microscopy for sarcoptes scabiei hominis. *Am. J. Med. Sci.* **2015**, *350*, 414. [CrossRef] [PubMed]
33. Cinotti, E.; Perrot, J.; Labeille, B.; Cambazard, F. Reflectance confocal microscopy for cutaneous infections and infestations. *J. Eur. Acad. Dermatol. Venereol.* **2016**, *30*, 754–763. [CrossRef] [PubMed]
34. Moher, D.; Liberati, A.; Tetzlaff, J.; Altman, D.G. Preferred reporting items for systematic reviews and meta-analyses: The prisma statement. *BMJ* **2009**, *339*, b2535. [CrossRef] [PubMed]
35. McGrath, T.A.; Alabousi, M.; Skidmore, B.; Korevaar, D.A.; Bossuyt, P.M.M.; Moher, D.; Thombs, B.; McInnes, M.D.F. Recommendations for reporting of systematic reviews and meta-analyses of diagnostic test accuracy: A systematic review. *Syst. Rev.* **2017**, *6*, 194. [CrossRef] [PubMed]
36. Whiting, P.F.; Rutjes, A.W.; Westwood, M.E.; Mallett, S.; Deeks, J.J.; Reitsma, J.B.; Leeflang, M.M.; Sterne, J.A.; Bossuyt, P.M. Quadas-2: A revised tool for the quality assessment of diagnostic accuracy studies. *Ann. Intern. Med.* **2011**, *155*, 529–536. [CrossRef] [PubMed]
37. Leeflang, M.M.G. Systematic reviews and meta-analyses of diagnostic test accuracy. *Clin. Microbiol. Infect.* **2014**, *20*, 105–113. [CrossRef] [PubMed]
38. Deeks, J.J.; Macaskill, P.; Irwig, L. The performance of tests of publication bias and other sample size effects in systematic reviews of diagnostic test accuracy was assessed. *J. Clin. Epidemiol.* **2005**, *58*, 882–893. [CrossRef] [PubMed]
39. Nori, S.; Rius-Díaz, F.; Cuevas, J.; Goldgeier, M.; Jaen, P.; Torres, A.; González, S. Sensitivity and specificity of reflectance-mode confocal microscopy for in vivo diagnosis of basal cell carcinoma: A multicenter study. *J. Am. Acad. Dermatol.* **2004**, *51*, 923–930. [CrossRef]
40. Gerger, A.; Koller, S.; Weger, W.; Richtig, E.; Kerl, H.; Samonigg, H.; Krippl, P.; Smolle, J. Sensitivity and specificity of confocal laser-scanning microscopy for in vivo diagnosis of malignant skin tumors. *Cancer* **2006**, *107*, 193–200. [CrossRef]
41. Guitera, P.; Menzies, S.W.; Longo, C.; Cesinaro, A.M.; Scolyer, R.A.; Pellacani, G. In vivo confocal microscopy for diagnosis of melanoma and basal cell carcinoma using a two-step method: Analysis of 710 consecutive clinically equivocal cases. *J. Invest. Dermatol.* **2012**, *132*, 2386–2394. [CrossRef] [PubMed]
42. Longo, C.; Farnetani, F.; Ciardo, S.; Cesinaro, A.M.; Moscarella, E.; Ponti, G.; Zalaudek, I.; Argenziano, G.; Pellacani, G. Is confocal microscopy a valuable tool in diagnosing nodular lesions? A study of 140 cases. *Br. J. Dermatol.* **2013**, *169*, 58–67. [CrossRef] [PubMed]
43. Peppelman, M.; Wolberink, E.A.; Blokx, W.A.; van de Kerkhof, P.C.; van Erp, P.E.; Gerritsen, M.J. In vivo diagnosis of basal cell carcinoma subtype by reflectance confocal microscopy. *Dermatology* **2013**, *227*, 255–262. [CrossRef] [PubMed]
44. Rao, B.K.; Mateus, R.; Wassef, C.; Pellacani, G. In vivo confocal microscopy in clinical practice: Comparison of bedside diagnostic accuracy of a trained physician and distant diagnosis of an expert reader. *J. Am. Acad. Dermatol.* **2013**, *69*, e295–e300. [CrossRef] [PubMed]

45. Pellacani, G.; Pepe, P.; Casari, A.; Longo, C. Reflectance confocal microscopy as a second-level examination in skin oncology improves diagnostic accuracy and saves unnecessary excisions: A longitudinal prospective study. *Br. J. Dermatol.* **2014**, *171*, 1044–1051. [CrossRef] [PubMed]
46. Castro, R.P.; Stephens, A.; Fraga-Braghiroli, N.A.; Oliviero, M.C.; Rezze, G.G.; Rabinovitz, H.; Scope, A. Accuracy of in vivo confocal microscopy for diagnosis of basal cell carcinoma: A comparative study between handheld and wide-probe confocal imaging. *J. Eur. Acad. Dermatol. Venereol.* **2015**, *29*, 1164–1169. [CrossRef]
47. Farnetani, F.; Scope, A.; Braun, R.P.; Gonzalez, S.; Guitera, P.; Malvehy, J.; Manfredini, M.; Marghoob, A.A.; Moscarella, E.; Oliviero, M. Skin cancer diagnosis with reflectance confocal microscopy: Reproducibility of feature recognition and accuracy of diagnosis. *JAMA Dermatol* **2015**, *151*, 1075–1080. [CrossRef]
48. Guitera, P.; Menzies, S.W.; Argenziano, G.; Longo, C.; Losi, A.; Drummond, M.; Scolyer, R.A.; Pellacani, G. Dermoscopy and in vivo confocal microscopy are complementary techniques for diagnosis of difficult amelanotic and light-coloured skin lesions. *Br. J. Dermatol.* **2016**, *175*, 1311–1319. [CrossRef]
49. Nelson, S.A.; Scope, A.; Rishpon, A.; Rabinovitz, H.S.; Oliviero, M.C.; Laman, S.D.; Cole, C.M.; Chang, Y.H.H.; Swanson, D.L. Accuracy and confidence in the clinical diagnosis of basal cell cancer using dermoscopy and reflex confocal microscopy. *Int. J. Dermatol.* **2016**, *55*, 1351–1356. [CrossRef]
50. Witkowski, A.; Łudzik, J.; DeCarvalho, N.; Ciardo, S.; Longo, C.; DiNardo, A.; Pellacani, G. Non-invasive diagnosis of pink basal cell carcinoma: How much can we rely on dermoscopy and reflectance confocal microscopy? *Skin Res. Technol.* **2016**, *22*, 230–237. [CrossRef]
51. Peccerillo, F.; Mandel, V.D.; Di Tullio, F.; Ciardo, S.; Chester, J.; Kaleci, S.; de Carvalho, N.; Del Duca, E.; Giannetti, L.; Mazzoni, L.; et al. Lesions mimicking melanoma at dermoscopy confirmed basal cell carcinoma: Evaluation with reflectance confocal microscopy. *Dermatology* **2018**, *235*, 35–44. [CrossRef] [PubMed]
52. Harbord, R.M.; Whiting, P. Metandi: Meta-analysis of diagnostic accuracy using hierarchical logistic regression. *Stata J.* **2009**, *9*, 211–229. [CrossRef]
53. Pellacani, G.; Vinceti, M.; Bassoli, S.; Braun, R.; Gonzalez, S.; Guitera, P.; Longo, C.; Marghoob, A.A.; Menzies, S.W.; Puig, S.; et al. Reflectance confocal microscopy and features of melanocytic lesions: An internet-based study of the reproducibility of terminologyfeatures of melanocytic lesions. *Arch. Dermatol.* **2009**, *145*, 1137–1143. [CrossRef] [PubMed]
54. Van Dijk, A.; den Outer, P.N.; Slaper, H. Climate and Ozone Change Effects on Ultraviolet Radiation and Risks (Coeur) Using and Validating Earth Observation. Available online: http://www.rivm.nl/dsresource?objectid=rivmp:9586&type=org&disposition=inline&ns_nc=1 (accessed on 6 July 2019).
55. Leiter, U.; Eigentler, T.; Garbe, C. Epidemiology of skin cancer. *Adv. Exp. Med. Biol.* **2014**, *810*, 120–140. [PubMed]
56. McKenzie, C.A.; Chen, A.C.; Choy, B.; Fernandez-Penas, P.; Damian, D.L.; Scolyer, R.A. Classification of high risk basal cell carcinoma subtypes: Experience of the ontrac study with proposed definitions and guidelines for pathological reporting. *Pathology* **2016**, *48*, 395–397. [CrossRef] [PubMed]
57. Trakatelli, M.; Morton, C.; Nagore, E.; Ulrich, C.; Del Marmol, V.; Peris, K.; Basset-Seguin, N. Update of the european guidelines for basal cell carcinoma management. *Eur. J. Dermatol.* **2014**, *24*, 312–329. [CrossRef] [PubMed]
58. Cohen, J.F.; Korevaar, D.A.; Altman, D.G.; Bruns, D.E.; Gatsonis, C.A.; Hooft, L.; Irwig, L.; Levine, D.; Reitsma, J.B.; de Vet, H.C.W.; et al. Stard 2015 guidelines for reporting diagnostic accuracy studies: Explanation and elaboration. *BMJ Open* **2016**, *6*, e012799. [CrossRef]

© 2019 by the authors. Licensee MDPI, Basel, Switzerland. This article is an open access article distributed under the terms and conditions of the Creative Commons Attribution (CC BY) license (http://creativecommons.org/licenses/by/4.0/).

Review

Deciphering the Molecular Landscape of Cutaneous Squamous Cell Carcinoma for Better Diagnosis and Treatment

Andreea D. Lazar [1], Sorina Dinescu [1,2,*] and Marieta Costache [1,2]

[1] Department of Biochemistry and Molecular Biology, University of Bucharest, 050095 Bucharest, Romania; andreea.lazar@bio.unibuc.ro (A.D.L.); marieta.costache@bio.unibuc.ro (M.C.)
[2] The Research Institute of the University of Bucharest, 050663 Bucharest, Romania
* Correspondence: sorina.dinescu@bio.unibuc.ro

Received: 26 June 2020; Accepted: 13 July 2020; Published: 14 July 2020

Abstract: Cutaneous squamous cell carcinoma (cSCC) is a common type of neoplasia, representing a terrible burden on patients' life and clinical management. Although it seldom metastasizes, and most cases can be effectively treated with surgical intervention, once metastatic cSCC displays considerable aggressiveness leading to the death of affected individuals. No consensus has been reached as to which features better characterize the aggressive behavior of cSCC, an achievement hindered by the high mutational burden caused by chronic ultraviolet light exposure. Even though some subtypes have been recognized as high risk variants, depending on certain tumor features, cSCC that are normally thought of as low risk could pose an increased danger to the patients. In light of this, specific genetic and epigenetic markers for cutaneous SCC, which could serve as reliable diagnostic markers and possible targets for novel treatment development, have been searched for. This review aims to give an overview of the mutational landscape of cSCC, pointing out established biomarkers, as well as novel candidates, and future possible molecular therapies for cSCC.

Keywords: cutaneous squamous cell carcinoma; ultraviolet radiation; genes; microRNAs; lncRNAs; novel therapeutic approaches

---

1. Introduction

Skin is the largest human organ and serves as the first line protective barrier against environmental assaults. Accumulation of these stresses (sun damage, microorganisms, noxious agents) can lead to cutaneous neoplasia, commonly named skin cancer. Cutaneous cancer represents the most common worldwide malignancy, and its incidence shows few signs of plateauing. It is generally divided into malignant melanoma and non-melanoma skin cancer (NMSC), the latter including basal cell carcinoma (BCC) and squamous cell carcinoma (SCC) as the major subtypes [1]. The impact of cutaneous cancer at a global level is vast, in the order of millions of cases every year, with patients being far more often diagnosed with either BCC or SCC, than with malignant melanoma. Annually, about 4.3 million new cases of BCC, 1 million cases of SCC and ~200,000 cases of melanoma are registered in the United States alone, with most cases found on sun-exposed areas of the body. Due to high-associated mortality, this specific cutaneous cancer is rightly perceived as much more deadly, when compared to NMSC. Nonetheless, NMSC cases are not to be trifled with, and represent a definitive cause for concern, with more than 5400 deaths worldwide each month [2,3]. When choosing ethnicity as a monitoring criteria, cutaneous cancer represents approximately 2–4% in Asians, 4–5% in Hispanics, and 1–2% in people of African descent, with SCC being the most common cutaneous cancer in the last group [4]. Numerous attempts have been made to reduce the number of cases, by informing the public about the risk factors involved in the appearance of cutaneous carcinoma (exposure to ultraviolet radiation, family history, genetic predisposition, light skin color, etc.), the means for prevention and the importance of early diagnosis, but still, the incidence continues to rise [5]. Thus, the personal, medical and financial issues associated with cutaneous carcinoma continue to represent a heavy burden on patients' life and clinical management [2,3].

Cutaneous SCC (cSCC), the second most common type of skin cancer, develops preferentially in the interfollicular epidermis, as a consequence of the unrestricted proliferation of epidermal keratinocytes. Its appearance is strongly associated with the development of precursor lesions, namely actinic keratoses, signs of chronic sun damage, which result from the proliferation of atypical epidermal keratinocytes. Most such precancerous lesions will not progress to cSCC tumorigenesis, but simply persist or may even regress. Even so, they pose an increased risk of neoplasia and should be treated accordingly [6]. cSCC is considered highly curable, because its metastatic rate is quite low (1–5% of cases) and surgical removal of the affected tissue is usually very effective in treating this form of cancer. This depends of course on the gravity of said cSCC, as once metastatic it usually displays a rapacious behavior [7]. No consensus has been reached as to which features better characterize the aggressiveness of cSCC. Some subtypes (adenosquamous, desmoplastic) have been recognized as high risk variants, but, depending on certain tumor features (size, location, depth, etc.), cSCC that are normally thought of as low risk could pose increased danger to the patients. As a result of this uncertainty, molecular markers have been searched for as reliable biomarkers for cSCC and possible targets for novel treatment development [8].

New findings with regards to the molecular patterns involved in neoplastic transformation of cells have come to light in the last decades. Modern techniques, such as next generation sequencing, have made it possible to highlight some important mutational markers. It is now well known that, once proto-oncogenes acquire mutations, and thus convert to oncogenes, cell growth and proliferation will be uncontrolled. Similarly, alterations in tumor suppressor genes, which have the function to inhibit cell growth, can easily lead to unregulated cell proliferation as a result of the loss of negative control. As such, any dysregulation of the proto-oncogenes and tumor suppressors represents the basic mechanism behind tumor development and growth [5]. Research for a pathway of similar significance to SCC as the Hedgehog signaling cascade for BCC, meaning that mutations appearing in that pathway lead to oncogenesis, is ongoing. Sequencing of the whole genome from cSCC revealed an intense mutational profile, with an average of one mutation per 30,000 base pairs [9]. This discovery has hindered the identification of key driver mutations. Another disparity between the two NMSC is that BCC apparently arises de novo, while SCC can develop from precursor lesions (in 65% cases from actinic keratosis). A genomic analysis of such precursor samples and cSCC specimens, revealed that the former display a lower mutational burden, thus suggesting an earlier stage of tumor evolution [10]. It was therefore concluded that the mutations acquired in cSCC, but not actinic keratosis, might be the specific mutations that drive progression from premalignant to malignant forms [11]. Mutations in several genes and pathways have been suggested to determine the development of cSCC [12], and several noncoding RNA molecules have been found to be abnormally expressed in this type of cancer [13,14].

This review aims to provide an understanding of the current knowledge regarding the genomic landscape of cSCC, pointing out relevant disease biomarkers and potential targets, which could facilitate the future diagnosis and treatment of cSCC. Firstly, we describe the major risk factor associated with the development of this type of cutaneous carcinoma and the means for prevention and early diagnosis. Then, we summarize the most frequently mutated genes associated with cSCC, as well as recently discovered ones. We continue with an up to date overview of the noncoding RNA modifications and finalize with a brief description of the current therapeutic options, as well as potentially new ones for cSCC.

2. Etiology, Prevention and Early Diagnosis of Cutaneous SCC

An overwhelming number of epidemiological and experimental investigations have deemed cumulative exposure to ultraviolet (UV) radiation as the main environmental risk factor for the pathogenesis of cSCC [2,15]. Apart from UV radiation, inherited genetic conditions (xeroderma pigmentosum, albinism, epidermolysis bullosa), human papillomavirus (HPV) infections, severe arsenic exposure, chronic immunosuppressed state (organ transplantation) or precancerous lesions (such as actinic keratosis) are recognized as predisposing factors in the development of cSCC [12,15–17]. Of the three subtypes of UV radiation (A, B, C, distinguished by wavelength), only UV-A and UV-B are considered clinically relevant for the pathogenesis of skin cancer, because UV-C is absorbed entirely by the atmosphere. The daily dosage of UV-B is much lower than UV-A, however, UV-B is far more dangerous, because it is strongly absorbed by the cellular nucleus DNA, and proteins in the epidermis, thus exerting its effect on the genetic material of epidermal keratinocytes, from which cSCC originates (Figure 1). UV-B is also responsible for the majority of sunburns. Upon stimulation by UV exposure, melanocytes from the basal layer of epidermis act to absorb UV by undergoing melanogenesis, in which they produce the photoprotective pigment melanin that is also distributed to keratinocytes. As a result of this, the incidence of skin cancer is much lower in individuals with darker skin phenotypes, which possess higher levels of photoprotective pigment [18]. However, the protection is prone to failure in case of repeated exposure to intense UV radiation, thus, cutaneous damage will appear, at first in the form of a sunburn. UV-B rays directly induce DNA lesions (misbonding of two pyrimidines, either thymine or cytosine, within the same DNA strand), because the wavelength of this specific radiation corresponds to the absorption spectrum of the genetic material. As such, UV-B photons are directly absorbed and lead to formation of cyclobutane pyrimidine dimers and pyrimidine 6-4 pyrimidone photoproducts, which left unrepaired become mutagenic [5,19]. In contrast to UV-B, the exact role of UV-A in cutaneous carcinogenesis is not clearly understood. For a long time, it was considered that UV-A has a minor role in skin cancer development, because the photons are not directly absorbed by DNA. However, researchers have discovered that UV-A causes indirect damage to DNA by the generation of reactive oxygen species, crosslinks between DNA and proteins and even the direct formation of single-strand DNA breaks or cyclobutane pyrimidine dimers [20]. Epidemiologic studies also seem to support these harmful effects, and it has been reported that a single indoor tanning session, during which UV-A radiation emission is substantially higher than from natural sun [21], can increase the risk of developing cSCC by 67% [22]. In light of these findings, alternative pathways that lead to skin carcinogenesis are currently being searched for, to understand the mechanisms behind UV-A induced mutations [5].

Due to the fact that UV radiation, regardless of its natural or artificial origin, is considered to be the main environmental risk factor in the etiology of cSCC, and also because, in early stages, cSCC can be cured with good prognosis, this type of cancer represents an ideal candidate for combating by means of primary and secondary (early detection) prevention [2]. The first steps in the prevention of cutaneous cancer are constantly informing and reminding the public about the dangers that come with exposure to the UV light. This can be carried out through huge promotion of sun creams and campaigns such as the 'Slip (on a shirt), Slop (on some sunscreen), Slap (on a hat)' initiative and the following SunSmart campaign in Australia, or the periods of live program (POLP) in Germany [23]. All these programs aim to provide the public with skincare routines that are concentrated on sun protection and exposure to UV radiation. Regular use of sunscreen with an SPF 15 or higher reduces the risk of developing skin cancer by approximatively 40% [24,25]. The next step, secondary prevention of cSCC, is achieved by early observations of the precancerous lesions, in order to identify the first stages of cancer, which, luckily, can be treated with the proper medicine and self-care. This step is reached with the help of public screening campaigns and the monitoring of skin cancer risk groups. For cSCC, the risk groups include patients with skin type I (white), patients who suffer from actinic keratosis (precancerous lesions) and patients who have been previously diagnosed with cSCC [15,26]. The aim of standardizing such groups is to establish a reliable set of prognostic biomarkers, as specific and as sensitive as possible. In addition to this, a set of molecular biomarkers is currently being searched for.

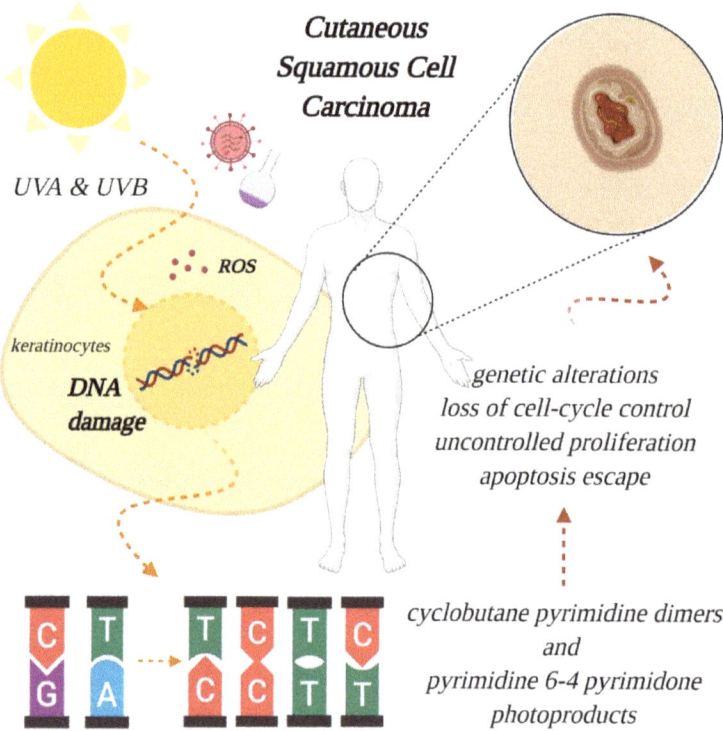

Figure 1. Effect of ultraviolet radiation on the genetic material of epidermal keratinocytes, from which cutaneous squamous cell carcinoma (SCC) originates. Excessive absorption of ultraviolet (UV) light generates oxidative stress, through formation of reactive oxygen species (ROS), and breaks the double helix, leading to aberrant binding of pyrimidines and further genetic alterations, culminating with tumor formation. Other risk factors include viral infections and chemical exposure (created in BioRender.com).

3. Established SCC-Associated Markers

Data from the genomic analyses (next generation sequencing (NGS) and whole exome sequencing (WES)) have identified some genes to be frequently mutated in cSCC, establishing them as driver genes. Apart from *TP53*, which is one of the first inactivated tumor suppressor genes, a handful of key mutations frequently found in SCC of the skin have been proposed, among them *CDKN2A*, *NOTCH 1*, *NOTCH 2*, *FAT1* and *RAS* family members, involved in different cellular processes, such as cell-cycle control, squamous cell differentiation, survival and proliferation (Figure 2). Frequency of mutations in cSCC-associated genes across published studies can be found in Table 1, and differences may be attributed to the detection method, number and heterogeneity of the evaluated samples [27–31].

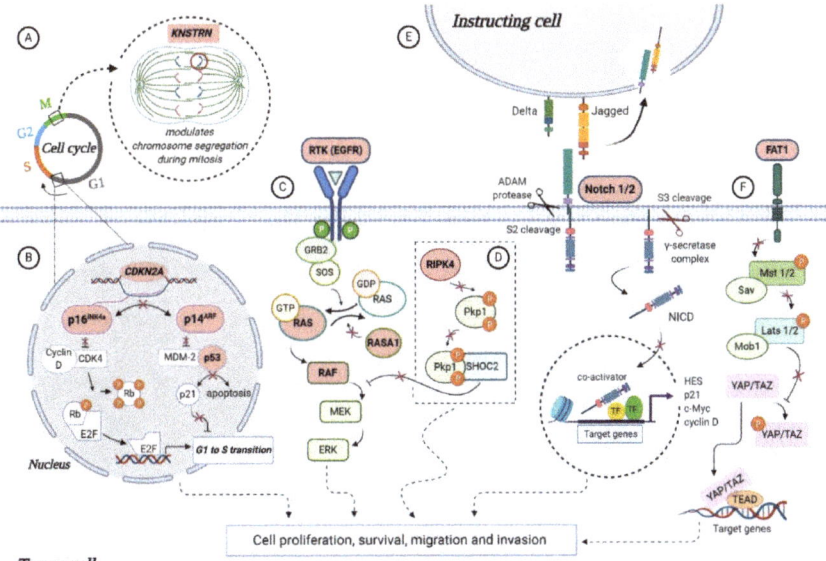

Figure 2. Molecular alterations that drive cutaneous squamous cell carcinoma (cSCC) proliferation, survival and metastasis through aberrant signaling (highlighted in pink): (**A**) alterations in *KNSTRN* expression promote abnormal chromosome segregation during mitosis; (**B**) *CDKN2A* encodes for cell-cycle regulatory proteins p16^{INK4A} and p14^{ARF}, involved in retinoblastoma (RB) and p53 pathways. loss of heterozygosity (LOH), mutations or deletions of *CDKN2A* leads to functional loss of: (i) p16^{INK4A}, which allows phosphorylation of RB by CDK4-Cyclin D complex and release of E2F transcription factors, that can then transcribe S phase promoting genes; (ii) p14^{ARF}, which allows MDM-2 to bind p53 and inhibit apoptosis; (**C**) activating mutations in *EGFR*, *RAS* and *RAF* or inactivation of negative regulator *RASA1* promotes cell proliferation and survival through constitutive activation of MAPK pathway; (**D**) proposed model for RIPK4 action in skin carcinogenesis that depicts the phosphorylation of PKP1 by RIPK4, which promotes binding to scaffold protein SHOC2 and blocking of RAS/MAPK signaling. In the absence of functional RIPK4, the complex cannot assemble and the signaling pathway remains active, thus facilitating cSCC development; (**E**) the inactive precursor is cleaved in the Golgi by a furin-like convertase (S1 cleavage) and translocated into the cell membrane, where binding of a NOTCH ligand (Delta, Jagged) to the receptor induces the second cleavage (S2) by a member of the disintegrin and metalloproteinases (ADAM) family. This results in the formation of a membrane-tethered NOTCH truncated fragment, which is further cleaved (S3) by a presenilin-dependent γ-secretase complex, generating the NOTCH intracellular domain (NICD). The active form of the NOTCH receptor (NICD) can now enter into the nucleus, where it exerts its transcriptional activity. Inactivation of NOTCH 1/2 favors cSCC progression, however, the specific functional significance of this mutation has yet to be described; (**F**) the molecular mechanisms that contribute to tumor development in the context of FAT1 functional loss are poorly understood in cSCC, however, a model proposed for HNSCC suggests FAT1 acts as a scaffold for Hippo kinases, favoring the activation of the complex and the phosphorylation of YAP, which is sequestered in the cytoplasm or degraded. Absence of FAT1 dismantles the Hippo core complex leading to YAP dephosphorylation and its translocation to the nucleus, where it interacts with TEAD to induce the expression of genes promoting tumor progression (created in BioRender.com).

3.1. TP53

A highly characterized gene in cSCC, the tumor suppressor gene *TP53* codes for the "Guardian of the Genome" protein p53, a critical regulator involved in various cellular activities, among them DNA repair, cell-cycle control and apoptosis [32]. In cSCC, mutations of p53 are frequent but atypical, as they do not appear within conserved regions, as in the case of other cancers. Instead, p53 alleles present UV signature mutations identified as 'hot spots' along their sequences, which cause the gene to become inactive and give rise to a p53 mutant protein, also inactive. The p53 alterations are primarily believed to bestow resistance to apoptosis upon the cells, in response to UV radiation (Figure 2B), thereby leading to positive selection of p53 mutant cells and clonal expansion [5]. Across different studies the mutational frequency of *TP53* ranges from 42% to ~95% (Table 1) [27–31,33], with a statistically higher rate of mutation in metastatic tumors relative to primary non-metastatic cSCC. Further studies are needed to understand the implications of this finding [30].

Table 1. Frequency of mutations in cSCC-associated genes across published studies.

Gene	No. of Analyzed Samples	Mutations (%)	References
Cell-cycle control			
TP53	100	42	[33]
	91	64	[27]
	39	94.9	[28]
	29	79	[29]
	28	54–85	[30]
	40	70	[31]
CDKN2A	100	28	[33]
	91	23	[27]
	39	43.6	[28]
	29	45	[29]
	28	29–42	[30]
	40	45	[31]
Keratinocyte differentiation			
NOTCH 1	100	54	[33]
	91	75	[27]
	39	59	[28]
	29	48	[29]
	28	50–63	[30]
	40	75	[31]
NOTCH 2	100	34	[33]
	91	63	[27]
	39	51.3	[28]
	29	31	[29]
	28	41–48	[30]
	40	50	[31]
FAT1	39	43.6	[28]
	170	40	[27]
	28	22–37	[30]
	40	60	[31]
RIPK4	39	28	[28]
	29	24	[29]
RAS signaling			
HRAS	100	6	[33]
	91	16	[27]
	39	20.5	[28]
	29	13	[29]
	28	12–13	[30]
	40	22.5	[31]
KRAS	91	13	[27]
	29	10	[29]

Table 1. *Cont.*

Gene	No. of Analyzed Samples	Mutations (%)	References
BRAF	39	17.9	[28]
	29	13	[29]
	28	5–13	[30]
RASA1	39	13	[28]
Chromatin segregation/remodeling			
KNSTRN	100	19	[33]
KMT2C	39	38.5	[28]
	28	36–43	[30]
KMT2D	39	69.2	[28]
	28	31–62	[30]

3.2. CDKN2A

CDKN2A maps to chromosome 9 and encodes for p16^{INK4a} and p14^{ARF} (also referred to as p16 and p14), two cell-cycle regulatory proteins involved in retinoblastoma (RB) and p53 pathways, respectively. Loss of heterozygosity (LOH), mutations or homozygous deletions of *CDKN2A* lead to loss of function (Figure 2B), and have been associated with the progression of cSCC from actinic keratosis [34,35]. In a study investigating the potential pathways important in metastatic cutaneous SCC, both primary and metastatic samples of cSCC were compared using WES and targeted-sequencing. An increased rate of *CDKN2A* mutation (42%) was observed in the metastatic tumors, when compared to primary cutaneous SCC (29%) [30]. Another study, searching to validate tumor drivers and therapeutic targets, found *CDKN2A* to be mutated in 45% of 40 primary cSCC (20 well-differentiated and 20 moderately/poorly differentiated tumors), from both immunosuppressed and immunocompetent patients, by employing whole-exome analyses [31]. Across the published studies, the frequency of *CDKN2A* alteration varies from 23% to 45% (Table 1) [27–31,33].

3.3. RAS Signaling Genes

Among the genes carrying activating mutations in cSCCs are members of the RAS family, which consists of small guanosine triphosphate proteins (GTPases), involved in cellular signal transduction. When RAS is "switched on" by incoming signals, it subsequently activates other proteins found downstream (e.g., BRAF), culminating with the expression of specific genes involved in cell growth, differentiation and survival. RAS mutations at gene level can lead to the synthesis of permanently functional proteins, an outcome that can cause unintended and overactive cell signaling, even in the absence of an incoming signal (Figure 2C) [36]. In a study conducted by Li et al., out of the 29 metastatic cSCC samples evaluated, the majority of the activating mutations affected genes in the RAS/RAF/MEK/ERK pathway, such as HRAS, KRAS, the downstream kinase BRAF and the epidermal growth factor receptor (EGFR). Aside from an activating mutation, *EGFR* was also significantly recurrently amplified, though only one sample had a high-level gain [29]. Overexpression of *EGFR* seems to be a common feature of SCC, and an early event in squamous carcinogenesis [37].

Gain-of-function mutations of *HRAS* have been identified in up to 23% of cSCC [27–31], with a higher incidence in patients treated with BRAF inhibitors. A targeted sequencing analysis of 21 cSCC samples collected from patients receiving the BRAF inhibitor vemurafenib identified activating *RAS* mutations in 60% of the tumor samples, with *HRAS* being more commonly affected than other members of the RAS family, indicating the potent effect of BRAF repression on the other signaling molecules involved in RAS/RAF/MEK/ERK pathway [38,39]. In general, *HRAS* mutation is more commonly associated with cSCC than *KRAS* (10–13%) and *NRAS* (5%) [27,29] (Table 1). Inman et al. identified oncogenic activating mutations in *HRAS* [31], which have previously been identified in 3–20% of cSCC [28,29]. Notably, 10% of the samples exhibited copy number loss of *HRAS*, a result others have observed as well [29], warranting the need for better understanding of the role of *HRAS* in cSCC [31]. Concerning *BRAF* alterations, their frequency differs between primary (5%) and metastatic tumors (13–18%) across published studies [28–30].

3.4. NOTCH Signaling Genes

Frequently affected in cSCC, *NOTCH1* and *NOTCH2* genes encode for the members of the NOTCH family of transmembrane receptors with the same name, and represent direct targets of the transcription factor p53. The NOTCH signaling pathway (Figure 2E) they patronize is crucial to epidermal development and maturation, contributing to keratinocyte differentiation, therefore, any changes in NOTCH activity could destabilize this process [40]. While NOTCH1 is expressed throughout the epidermis, NOTCH2 is localized primarily in the basal layer [41]. Inactivation of *NOTCH1* and *NOTCH2* through point mutations in functional domains or truncation mutations have been identified in up to 75% for *NOTCH1* and 63% for *NOTCH2*, through WES of cutaneous SCC samples [27,31] (Table 1). The mutation of *NOTCH1* is considered an early event in squamous carcinogenesis of the skin, and its loss is associated with disease progression. In their study, South et al. presented a comprehensive mutation analysis of NOTCH1 in 130 samples of cSCC and squamoproliferative lesions, plus 10 matched, normal skin samples, using exome-level sequencing and validation by targeted deep sequencing. They demonstrated that NOTCH1 receptor is significantly mutated in 75% of sporadic cSCCs ($n = 91$), 49% of squamoproliferative lesions arising in patients receiving vemurafenib ($n = 39$) and 70% of normal skin samples ($n = 10$, four perilesional and six separate from lesion), thus confirming NOTCH1 receptor mutations as an early event and major tumor suppressor mechanism in carcinogenesis of cSCC [27].

3.5. FAT1

This gene encodes for the cadherin-*like* protein tumor suppressor FAT atypical cadherin 1, a member of the cadherin superfamily involved in the differentiation process of epidermal keratinocytes. Mutations of *FAT1* in cutaneous SCC are common, and range from 22% to 60% across different studies (Table 1), which searched to identify and validate driver genes and novel therapeutic targets using WES and targeted-sequencing [27,28,30,31]. *FAT1* was found to harbor nonsense mutations in 40–45% of both sporadic and aggressive cases of cSCC, leading to its inactivation [27,28], while another study that concentrated on the differential expression between primary and metastatic tumors found an increased rate of *FAT1* mutation in primary tumor samples (37%), in comparison to metastatic cSCC (22%) [30]. One study that focused on primary cSCCs from immunosuppressed and immunocompetent patients found *FAT1* to be mutated in 60% of the tumor samples [31]. While frequently encountered, the molecular mechanisms that contribute to tumor development in the context of FAT1 functional loss are poorly understood in cSCC. A proposed model in head and neck SCC (HNSCC) (Figure 2F) suggests that FAT1 acts as a scaffold for Hippo kinases, favoring the activation of the complex and the phosphorylation of Yes-associated protein (YAP), which is sequestered in the cytoplasm or degraded. Absence of FAT1 dismantles the Hippo core complex, leading to YAP dephosphorylation and its translocation to the nucleus, where it interacts with TEAD to induce the expression of genes promoting tumor progression [42].

4. Novel SCC-Associated Markers

Additional improvements in genomic analysis techniques have led to the identification of novel genes that could drive cSCC development, shedding further light on the vast mutational landscape of this specific skin cancer. Alterations of genes involved in keratinocyte differentiation, RAS signaling, chromatin segregation and remodeling, as well as other potential cSCC-associated genes, have been found in independent studies (Table 1), although a consensus on reliable novel driver genes has not been reached. However, it is important to mention that discrepancies across different studies concerning the list of novel key mutations probably reflect the clinico-pathological heterogeneity of cSCC analyzed samples, and the employed technique for detection [28–31,33].

4.1. KNSTRN

KNSTRN gene encodes a kinetochore associated protein, with the function to modulate onset of anaphase and segregation of chromosomes during mitosis. Point mutations at codon 24 of KNSTRN (UV signature mutations) have been observed in 19% of cSCC cases and 13% of precancerous lesions [33]. This affects the function of KNSTRN protein, and results in the disruption of chromatid cohesion in normal cells, an event that can lead to chromosomal aberrations or aneuploidy (Figure 2A). Studies to clarify its clinical applicability are needed of course, but mutations of this protein rarely occur in other malignancies, thus, it may represent a previously unidentified oncogene and a specific biomarker for cutaneous tumorigenesis [5,43].

4.2. RASA1

Another interesting candidate tumor suppressor gene in aggressive cSCC is *RAS p21 protein activator 1 (RASA1)*, found mutated in 13% of analyzed cases, with 66% of its mutations predicted to truncate or eliminate the protein [28]. Due to its high inactivation mutation ratio, it has also been identified as a candidate tumor suppressor gene in HNSCC [44]. The p120-RasGAP protein it encodes belongs to a family of RAS GTPase activating proteins (GAP), and functions as a negative regulator of pro-oncogenic RAS (Figure 2C), thus preventing cancer formation, although its exact role is not fully understood [45]. Inactivation of RASA1 and other members of the family through genomic loss, mutation or epigenetic silencing has been proposed to explain activation of the RAS signaling pathway in tumors that do not harbor specific RAS mutations. Despite the fact that *RASA1* is frequently inactivated by mutation in many other tumor types, its role in cSCC and cancer in general remains unclear [46].

4.3. RIPK4

This gene encodes for a serine/threonine kinase essential for squamous epithelial differentiation regulation [47], which has previously been reported as recurrently mutated in HNSCC [48]. Inactivating mutations in *RIPK4* are associated with popliteal pterygium syndrome, a severe autosomal recessive disease that affects the human face, limbs and genitalia [49]. In mice, a similar neonatal lethal syndrome is generated after knockout of RIPK4, which is accompanied by defective epidermal differentiation, including keratinocyte hyperplasia with expanded spinous and granular layers [47]. Pickering et al. identified this novel candidate driver gene of cSCC mutated in 28% of the tumors with a UV signature, with all mutations clustering in either exon 2 or exon 8, which encode the kinase and ankyrin repeat domains, respectively. They also observed a high ratio of nonsense, frameshift and splice mutations (35%), suggesting that a selection for inactivation of *RIPK4* occurs in cSCC. The clustering of mutations within the kinase and ankyrin repeat domains strongly indicated non-random mutations and supported the hypothesis that RIPK4 is a putative tumor suppressor for aggressive cSCC [28]. Li et al. arrived to the same conclusion when they found *RIPK4* recurrently altered in their cSCC cohort, with mutations in seven out of 29 samples (24%), and two of these mutations truncated, suggesting a recurrent inactivation of the gene [29]. Despite the potential significance of RIPK4 in cSCC, little is known about how it functions to regulate epidermal differentiation and tumorigenesis at the molecular level. A proposed model for RIPK4 action in skin carcinogenesis depicts the phosphorylation of desmosome protein plakophilin-1 (PKP1) by RIPK4, which promotes binding to scaffold protein SHOC2 and the blocking of RAS/MAPK signaling [50], illustrated in Figure 2D.

4.4. Chromatin Remodeling Genes

Genes important in chromatin remodeling and histone modification, such as *KMT2C* and *KMT2D*, showed high rates of mutations in several cSCC cases [28,30]. A study concerning identification of novel driver genes and therapeutic targets in aggressive cSCC found frequent inactivating mutations (~39%) in *KMT2C*, a gene which encodes a component of a histone methylation complex involved in transcriptional regulation. Their presence was correlated with significant shorter recurrent free survival for the patients, which were prone to faster recurrence and bone invasion [28]. Another study reported mutations in this gene in both primary cSCC (36%) and metastatic samples (43%), with a higher incidence in the latter [30]. Such mutations of *KMT2C* have also been reported for other types of tumors, including breast, bladder and gastric cancers, with reduced overall survival for the patients, as observed in the TCGA cancer datasets [51–53].

KMT2D, a histone methyltransferase that regulates H3 lysine 4, was also strongly mutated in 69% out of 39 aggressive cSCC samples analyzed through exome and targeted sequencing for identification of novel potential driver mutations [28], and in vitro studies have shown that *KMT2D* mutated cells display genomic instability and increased transcriptional stress [54]. In a study evaluating the differential mutation frequencies in metastatic cSCC versus primary tumors, only *KMT2D* showed significantly higher rates of mutation in the metastatic samples (62%) relative to non-metastatic ones (31%), implying a potential role in the development of cSCC aggressive behavior [30]. *KMT2D* alterations have also been reported in HNSCC (11–16%), esophageal SCC (14–19%) and cutaneous melanoma (19–29%), suggesting that common epigenetic pathways drive squamous cell carcinogenesis [54–56]. These provided data supports the two epigenetic regulators as potential new biomarkers for cSCC, driving this type of skin cancer towards aggressive behavior and poor outcome.

4.5. Other Potential Cutaneous SCC-Associated Genes

A recent analysis of the complex mutational landscape of cSCC, associated with the development of poorly differentiated and well-differentiated tumors in both immunosuppressed and immunocompetent patients, has identified several new potential driver genes correlated with tumor development [31]. The analysis implicates *SEMA3C, STEAP4, MMP10, RAP2B* and *AP2M1* as potential cSCC drivers, genes with known implications in other types of carcinoma. Semaphorin-3C (SEMA3C) promotes prostate cancer growth by transactivating multiple receptor tyrosine kinases (RTK), including EGFR, via Plexin B1 receptor which has intrinsic GAP activity [57]. Furthermore, the overexpression of *SEMA3C* is associated with unfavorable outcomes in a wide spectrum of tumors, including glioma, breast, lung, liver, pancreatic, stomach and gynecological cancers [58]. *STEAP4* encodes for a member of the six transmembrane epithelial antigen of prostate (STEAP) family, which functions as a metalloreductase and may promote prostate and colorectal cancer development [59,60]. Stromelysin-2, also known as matrix metalloproteinase-10 (MMP10), may mediate c-Fos driven cSCC development, and has been linked to lung cancer stem cell maintenance, tumor initiation and metastatic potential [61,62]. The intronless gene *RAP2B* is a well described oncogenic activator, belonging to the RAS-related family [63] and AP2M1, a component of the adaptor protein complex 2 (AP-2), may regulate senescence escape in response to chemotherapy through interaction with CTLA-4 immune checkpoint [64]. Furthermore, the analyses revealed some genes that may pre-dispose patients to well-differentiated tumors (alteration in sodium/potassium transporter ATP1A1) or poorly differentiated ones (alterations in Grainyhead like transcription factor 2 (GRHL2) and arginine methyltransferase PRMT3) [31]. Overall, these observations lend support to the hypothesis that the recent integrated analysis approach has potentially revealed novel drivers of cSCC, and provides further incentive for functional interrogation of the genes and pathways revealed in the study [31].

5. Non-Coding RNA Modifications in Cutaneous SCC

MicroRNAs (miRNAs/miRs) are non-coding transcripts of about 19–25 nucleotides in length, which regulate gene expression at a post-transcriptional level, by either causing mRNA degradation or blocking translation. In cancer, miRNAs can function as tumor suppressors or oncogenic miRNAs (onco-miRs), depending on the pathway in which they are involved [65,66]. While miRNAs have been heavily studied, and are well understood for their function in gene regulation, long non-coding

RNAs (lncRNAs) are less understood. LncRNAs are transcripts longer than 200 nucleotides, without open reading frames, that can interact with DNA, RNA or proteins to regulate gene expression via various pathways [67], and were found to play an active role in carcinogenesis [68]. Dysregulation of miRNAs and lncRNAs' expression has been shown to impact cell proliferation, resistance to apoptosis, the induction of angiogenesis, the promotion of metastasis and the evasion of tumor suppressors during cSCC development [13,14] (Figure 3); but their functions and molecular mechanisms still remain underexplored.

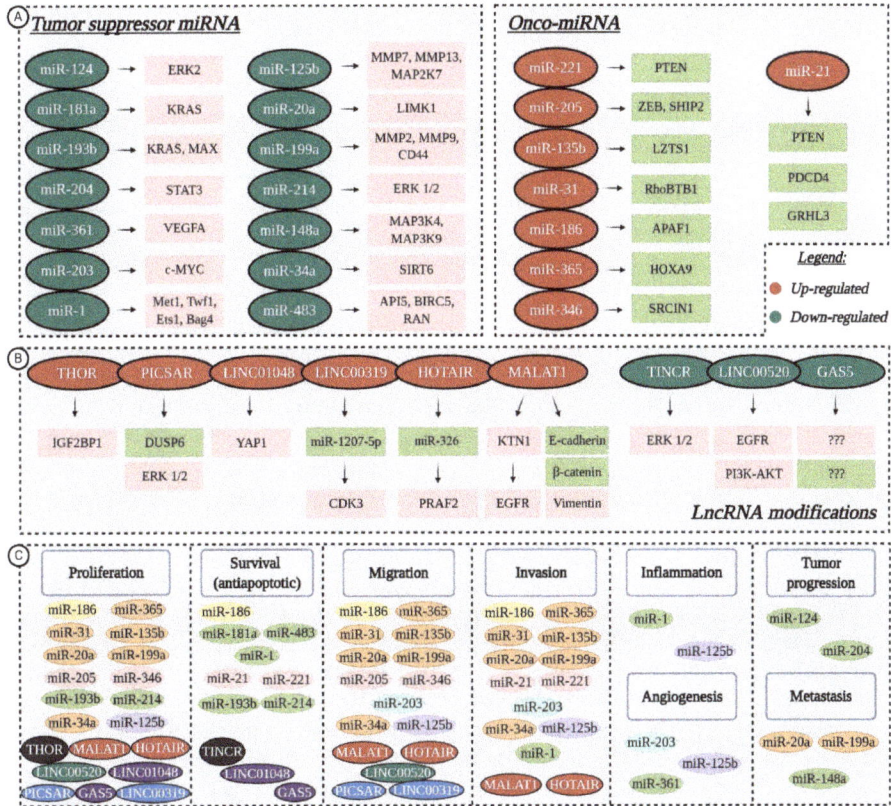

Figure 3. Non-coding RNA modifications and their functional roles in cSCC development: (**A**) down-regulation of tumor suppressor microRNAs (miRNAs) leads to the overexpression of target genes, while up-regulation of onco-miRNAs has a suppressive effect on specific molecular targets; (**B**) Dysregulated long non-coding RNAs (lncRNAs) that contribute to tumor progression through abnormal regulation of targeted genes (currently unknown for GAS5); (**C**) LncRNAs and miRNAs favor cell proliferation, survival, migration, invasion, inflammation, angiogenesis, tumor progression and metastasis in cSCC. Color-coded for involvement in several processes (created in BioRender.com).

5.1. Tumor Suppressor miRNAs Downregulated in cSCC

MicroRNA profiling studies showcased the altered expression of tumor suppressor miRNAs, while further research revealed their molecular targets and possible roles in cSCC evolution (Figure 3A). For instance, downregulation of miR-124 and miR-214 mediates tumor progression through the induction of ERK kinases that contribute to the MAPK signaling pathway, essential for cell proliferation, differentiation and survival. While miR-124 downregulation only affects ERK2, transfection of miR-214 mimic lowers the expression of both ERK1 and ERK2, thus establishing the two as targets of miR-214 [69,70]. Downregulation of miR-204 also contributes to malignant progression via the MAPK pathway modulation, by activating STAT3, which acts as a transcription factor when translocated into the nucleus, promoting tumor development [71]. The overexpression of miR-204 could inhibit STAT3 activation and translocation into the nucleus, with consequent inhibition of carcinoma progression [72].

Compared to healthy skin, the expression of miR-193b/365a cluster was significantly altered in a mouse model of two-stage chemically induced cSCC. The cluster exhibited decreased expression during tumor progression and was found to target KRAS and MAX, thus proving that miR-193b/365a act as synergistic co-regulators of the MAPK pathway, promoting cell proliferation and survival [73]. Reduced levels of miR-181a, which also targets KRAS seem to be essential for keratinocytes' transition towards cSCC, facilitating cell survival through continued MAPK signaling [74]. Upstream activators MAP3K4 and MAP3K9 are up-regulated in the absence of miR-148a, which, in turn, promotes proliferation and tumor metastasis [75].

Concerning tumor cell survival, the loss of miR-483-3p leads to overexpression of various anti-apoptotic genes, such as *API5*, *BIRC5* (also termed *Survivin*) and *RAS-related nuclear protein (RAN)*. In vivo intra-tumoral delivery of miR-483-3p has been shown to inhibit growth of cSCC xenografts, promoting it as a potential treatment [76]. Functional studies have also shown miR-1 to be involved in promoting cell survival, as well as invasion and inflammation, if down-regulated. As a result of its low expression, various target genes, among them, *Met*, *Twf1*, *Ets1* and *Bag4*, are overexpressed, causing pro-oncogenic changes in squamous epithelial cells, such as high secretion of MMPs, epidermal growth factor ligands, inflammatory mediators and the inhibition of terminal differentiation [77,78].

The underexpression of miR-34a is associated with the aggressive progression of cSCC [79,80]. Studies suggest that miR-34a is a tumor suppressor whose restoration inhibits proliferation, migration and invasion of cancer cells by modulating the expression of HMGB1 and SIRT6 [80]. The first target is a nuclear-binding protein that participates in the regulation of DNA organization and gene transcription, while the second targeted gene is a NAD+-dependent histone deacetylase and ADP ribosyl transferase that has been implicated in DNA repair, genomic stability and telomere structure [81]. The suppressive function of miR-34a also relies on its involvement in keratinocyte differentiation [79]. Techniques that modulate miR-34a expression could provide a starting point for valuable therapeutic tools. MRX34, which restores the function of endogenous miR-34, has already been tested in a clinical setting with promising results [82,83]. Pronounced angiogenesis is promoted by miR-203 and miR-361 in cSCC tumors compared to normal skin. Low levels of miR-361-5p induce VEGFA expression, while miRNA-203 was shown to exert its function, both in vitro and in vivo studies, by targeting the proto-oncogene c-MYC, and at the same time, facilitating cell migration and invasive growth [84,85].

Recent data has revealed miRNA molecules that take part in almost every stage of cSCC carcinogenesis, such as miR-20a and miR-199a whose down-regulation favors proliferation, migration, invasion and metastasis, or miR-125b, which also partakes in inflammation and angiogenesis [86–88]. The expression of LIMK1, a known tumor metastasis promoter, is significantly higher in the absence of miR-20a, resulting in the inactivation of substrate cofilin, with subsequent formation of stress fibers and cell invasion [86]. In cSCC cell lines, miR-199a targeted CD44 to repress the proliferation, migration and invasion of tumor cells, and regulated the interaction between CD44 and Ezrin, a complex involved in metastasis [87,89]. A non-kinase transmembrane proteoglycan, CD44 exerts its effects on tumor cells by modulating cytoskeletal architecture and activating various protein kinases or transcription factors [90]. Apart from CD44, the down-regulation of miR-199a increased the activity of matrix metallopeptidases MMP2 and MMP9, important for epithelial to mesenchymal transition (EMT) [87]. In the A431 and UT-SCC-7 cell lines, the absence of miR-125b stimulates tumor cell growth, migration, invasion, inflammation and angiogenesis, apparently by targeting MMP7, MMP13 and MAP2K7, as discovered through bioinformatic analyses [88].

5.2. Oncogenic miRNAs Upregulated in cSCC

Substantial progress has been made in the past few years in identifying the target genes and functional roles of several onco-miRNAs linked to cSCC development (Figure 3A), which could serve as new therapeutic biomarkers for this type of cutaneous cancer [13,91]. Gong et al. demonstrated that miR-221 is significantly higher in cSCC tissues and cell lines than in normal samples, and it can operate as an oncomir [92]. Functional experiments showed that knockdown of miR-221 inhibited cell cycle and proliferation, while the upregulation of said miRNA presented the opposite effect. PTEN was identified as a direct target gene of miR-221. After transfection with miR-221 mimics, the dual reporter gene assays showed decreased levels of PTEN mRNA and protein expression, which induces the activation of the PI3K/AKT/mTOR pathway, hence, promoting the survival and invasion of tumor cells [92]. Similarly, in A431 cell line, miR-21 downregulates the expression of PTEN, and another tumor suppressor, PDCD4, promoting cell survival and invasion [93]. Targeting of the tumor suppressor GRHL3 by a miR-21-dependent network also results in PTEN loss, and the induction of aggressive cSCC [94]. Moreover, in immunocompromised patients and organ transplant recipients, cancer survival and invasion are favored by the up-regulation of miR-135b [95]. This specific miRNA modulates LZTS1, a tumor suppressor critical for normal mitosis progression, whose absence impairs Cdk1/Cdc25C interaction during the M phase and shortens the mitotic division time, causing improper chromosome segregation [95,96].

The miR-346-induced proliferation and migration of A431 cells is caused by the downregulation of SRCIN1. Data from the luciferase reporter assay indicated that SRCIN1 as a direct target gene of miR-346, via the 3′-UTR. SRCIN1 protein and mRNA levels, was suppressed, due to the ectopic expression of miR-346, which, in turn, facilitated cell proliferation and migration. Further rescue experiments demonstrated that overexpression of SRCIN1 reduced the effects of miR-346 on A431 cells [97]. Upregulation of miR-205 also induced cancerous keratinocyte proliferation and migration by targeting lipid phosphatase SHIP2 [98,99]. In the same cell line, miR-186 targets APAF1, a key molecule in the intrinsic apoptosis pathway [100]. In response to cytochrome c release, APAF1 oligomerizes and forms the apoptosome [101], therefore, its downregulation as a consequence of miR-186 overexpression inhibits tumor cell apoptosis and promotes cSCC proliferation, invasion and migration [100]. At the same time, miR-31 favors the enhanced proliferation, motility and colony-forming ability of cSCC cell lines. Experiments concerning silencing by siRNA or knockdown in UT-SCC-7 and A431 cells showed that loss of miR-31 suppresses these processes, by directly targeting RhoBTB1, a member of the Rho family of small GTPases [102,103]. Finally, Zhou et al. identified HOXA9, a direct target of onco-miR-365, to be significantly downregulated in cSCC tumors and cell lines. Absence of HOXA9 positively regulates HIF-1α and its downstream glycolytic regulators, which contributes to the enhanced glycolysis in cSCC development, as further cell proliferation, migration and invasion [104].

5.3. Aberrant Activity of lncRNAs

Currently a hot topic in the field of cancer research [105,106], several studies have outlined the aberrant expression of lncRNAs in cSCC development (Figure 3B). For instance, Zhang et al. proposed a novel c-MYC-assisted MALAT1-KTN1-EGFR axis, which contributes to cSCC progression, and may serve as a new target for therapy. Metastasis associated lung adenocarcinoma transcript 1 (MALAT1) was found to regulate the protein expression of EGFR, but did not affect its mRNA expression. Transcriptomic sequencing identified kinectin 1 (KTN1) as the key mediator for the MALAT1 regulation of EGFR. Mechanistic studies revealed that MALAT1 interacts with c-MYC to form a complex, which directly binds to the promoter region of the KTN1 gene and enhances its transactivation to positively regulate EGFR protein expression, leading to increased cell proliferation [107]. The knockout of *MALAT1* decreased the protein expression of vimentin and increased E-cadherin and β-catenin, favoring cell migration and invasion [108]. Yu et al. found another well-known lncRNA, specifically HOTAIR, exhibited an obvious elevation in cSCC cell lines A431 and SCL-1 [109]. HOTAIR is widely involved in the regulation of tumor cell proliferation, apoptosis, angiogenesis, invasion and metastasis [110]. In cSCC, the increased expression of HOTAIR facilitated cell migration, proliferation and EMT, while its down-regulation impeded these malignant processes. Furthermore, HOTAIR competitively bound to miR-326, so as to positively modulate its expression and regulate *prenylated Rab acceptor 1 domain family, member 2 (PRAF2)* expression [109]. Liu et al. detected increased levels of THOR, a highly conserved lncRNA, mainly expressed in normal testis and tumors [111], in A431 cells. The knockdown of THOR downregulated IGF2BP1-dependent mRNAs, suppressing cell survival and proliferation. As such, targeting IGF2BP1 through THOR silencing might be a novel strategy for cSCC inhibition [112]. Piipponen et al. employed whole-transcriptome and RNA in situ hybridization analyses, succeeding in finding high levels of P38 inhibited cutaneous squamous cell carcinoma associated lincRNA (PICSAR) expression in cSCC cells. According to their study, PICSAR targeted dual specificity phosphate 6 (DUSP6), a negative regulator of ERK2 and enhanced MAPK/ERK signaling cascade. Functional studies revealed that PICSAR promotes in vitro cell proliferation and migration, as well as growth of human cSCC xenografts in vivo [113]. Another report detected intergenic length non-protein coding RNA 1048 (LINC01048) to be highly expressed in cSCC tissues and recurrence tissues, compared to adjacent normal and non-recurrence samples. The knockdown of LINC01048 led to the activation of the Hippo pathway through upregulation of YAP1. Further mechanism investigation revealed that LINC01048 increased the binding of TAF15 to YAP1 promoter to transcriptionally activate YAP1 in tumor cells. Finally, rescue assays demonstrated that YAP1 positive regulation by LINC01048 mediated cell proliferation and survival [114]. Li et al. reported the significant upregulation of LINC00319, a recently discovered cancer-related lncRNA transcribed from the intergenic region of chromosome 21, in cSCC tissues and cell lines. The increased expression of LINC00319 was associated with larger tumor size and lymphovascular invasion. Gain-of-function and loss-of-function approaches demonstrated that LINC00319 promoted tumor cell proliferation, accelerated cell cycle progression, facilitated migration and invasion, and inhibited apoptosis. Mechanistic studies revealed that LINC00319 exerts its oncogenic functions via miR-1207-5p-mediated regulation of *cyclin-dependent kinase 3 (CDK3)* in A431 cells. Taken together, the data implies a potential link between upregulation of LINC00319 and poor prognosis of cSCC [115].

Located on chromosome 19, the gene of terminal differentiation-induced ncRNA (TINCR) can promote epidermal differentiation through post-transcriptional mechanism. In this regard, the downregulation of TINCR in cSCC specimens could be correlated to the decrease in differentiation [116]. Additionally, some suggested that TINCR is involved in A431 cell apoptosis and autophagy induced by the combined treatment with 5-aminolevulinic acid and photodynamic therapy, via the ERK1/2-SP3 pathway [117]. Another lncRNA whose expression is lowered in cSCC is LINC00520, a new type that has only been reported in a few tumors. In A431 cells, LINC00520 targeted EGFR, thus inhibiting the PI3K-AKT signaling pathway and suppressing cell proliferation and migration. Consequently, loss of LINC00520 had the opposite effect on A431 cells [118]. Finally, significantly decreased expression of GAS5, a tumor suppressor usually induced by stress (e.g., cell-to-cell contact inhibition, serum deficiency), was observed in cSCC tissue samples, in contrast to normal skin [119,120]. Studies done on A431 cells determined that GAS5 promoted the proliferation and survival of tumor cells, although its molecular targets are currently unknown [121].

The aforementioned findings suggest that the aberrant expression of ncRNAs (low levels of tumor suppressors and overexpression of onco-promoters), as well as subsequent target genes' dysregulation, may be potential predictor biomarkers of cSCC outcome, and support them as putative targets for cSCC, with prospective therapeutic value.

6. Novel Therapeutic Approaches for Cutaneous SCC

The standard treatment for cSCC is represented by surgical resection of the affected tissue and the immediate area around the lesion, with various surgical modalities (standard excision, Mohs' micrographic surgery, curettage and electrodessication or cryosurgery), followed by chemotherapy or radiotherapy for patients with high-risk tumors, such as those who experience local recurrence or metastases. However, this therapeutic method generates lesions of different sizes and depths, which can be difficult to heal, while follow-up treatment has a systemic effect, instead of targeting the specific affected area, thus, weakening the patients' immune system without guaranteeing full efficiency. Furthermore, regenerative proliferation associated with chronic inflammation and oxidative stress during wound healing has been shown to contribute to skin tumor promotion [122]. As such, novel therapeutic approaches are being searched for, to overcome the current limitations and provide high-risk patients with efficient therapeutic alternatives, potentially increasing their chance of survival and decreasing the heavy financial and emotional burden.

6.1. Targeted Therapy

A significant progress in the treatment of cSCC is represented by the introduction of targeted therapy drugs, such as EGFR inhibitors. Overexpression of this growth factor receptor involved in RAS signaling is quite common in cSCC, thus, mapping it as a promising target for molecular therapy. Cetuximab, an inhibitor of EGFR has been developed and tested on high-risk cSCC patients in clinical trials, with positive results. A good outcome was reported for patients with locally advanced or regional SCC, while, for distant metastatic sites, it has remained inefficient [123–125]. Tyrosine kinase inhibitors have also been used to disrupt EGFR pathways in cSCC cases. Clinical studies on gefetinib and imatinib have yielded slightly positive responses, with modest antitumor activity in recurrent or metastatic cSCC, but with limited adverse effects [126,127]. Cetuximab has already been approved by the FDA for treatment of HNSCC, as a stand-alone treatment or in combination with conventional therapies for enhanced efficiency. Radiation therapy synergizes with cetuximab by inducing apoptosis and blocking secondary repair mechanisms, and studies have shown that in combination with chemotherapy EGFR inhibitors are efficient against metastatic cSCC [8,128,129].

6.2. Immunotherapy

Cutaneous SCC harbors a heavy mutational burden caused by UV radiation [9], increasing the likelihood of response to immunotherapy, with promising results being reported in clinical studies for use of checkpoint inhibitors in advanced cSCC [130]. Recently, human monoclonal antibody cemiplimab, that targets PD-1, has been approved by the FDA for patients with locally advanced or metastatic cSCC, unfit for curative surgery or radiation therapy [131]. While efficient in ~50% of aggressive cSCC cases, common adverse effects (rash, fatigue, diarrhea), as well as serious immune-mediated reactions, such as pneumonitis, colitis, hepatitis, nephritis, were reported [131], advising caution to be employed, especially for immunocompromised patients. Research is ongoing for the further development of immunotherapy drugs, with the consensus that checkpoint inhibitors will play a great role in cSCC treatment in the future.

6.3. Topical Treatment

Although not currently recommended for treating cSCC, case reports have shown promising results for topical imiquimod or 5-fluorouracil treatment, either alone or in combination [132]. Recently, Fayne et al. reported a case of biopsy-proven invasive cSCC in an elder Caucasian male patient, with a history of multiple actinic keratoses and no previous skin cancers, who declined surgical treatment due to cosmetic outcome concerns. A combination of topical 5% imiquimod cream, 2% 5-fluorouracil solution, and 0.1% tretinoin cream was used five nights/week under occlusion, for a treatment goal of 30 total applications. The patient was evaluated in clinic every two weeks, during which, the affected site was briefly treated with cryotherapy. Out of the 30 desired applications, the patient completed only 24, due to the burning pain associated with the treatment, however, the follow-up biopsy 15 months after completing the topical procedure revealed a dermal scar with no evidence of residual carcinoma. Therefore, the combination therapy of topical imiquimod, tretinoin and 5-fluorouracil application, coupled with intermittent cryotherapy, proved to be efficient in treating a small, invasive cSCC in this particular case. Nonetheless, prospective randomized-controlled clinical trials are warranted [133].

7. Discussion

cSCC tumors are heterogeneous and characterized by inherent evolution, propelled by genetic instability, which challenges diagnostics and complicates the development of targeted therapies. While monotherapies, such as EGFR inhibitors, may prove temporarily successful for patients with locally advanced or regional cSCC, they remain inefficient for metastatic sites [123–125], possibly because they are not radical enough. For instance, a much higher incidence of activating RAS mutations was detected in patients treated with vemurafenib, which hindered the intended repression of mitogen-activated protein kinase (MAPK) signaling pathway (involved in cell proliferation and survival), and rendered the BRAF inhibitor inadequate for cSCC treatment [38,39]. Consequently, targeting a single molecular signature is unlikely to combat the aggressive behavior of cSCC and yield the desired outcome for the patients, prompting researchers to try and find reliable combinations instead.

In this regard, the pivotal signaling routes for cSCC progression could serve as a starting point, by identifying and targeting multiple co-regulators at once. For example, the constitutive activation of RAS signaling pathways is favored by the aberrant expression of both genes and ncRNAs [13,14,27–31]. Activating mutations in EGFR [29,37], RAS [27–31,38,39] and RAF [28–30], as well as the inactivation of negative regulators RASA1 [45,46] and RIPK4 [50], promotes RAS signaling and facilitates cell proliferation and survival. The downregulation of tumor suppressors miR-124 and miR-214 mediates cSCC progression through induction of ERK kinases [69,70], while the reduced expression of miR-204 also targets MAPK cascade via STAT3 [71]. At the same time, the decreased expression of miR-181a and miR-193b/miR-365a cluster, which target KRAS, promotes continued MAPK signaling [73,74]. Moreover, in the absence of miR-148a, upstream activators MAP3K4 and MAP3K9 are up-regulated, again favoring RAS signaling [75]. Concerning the action of lncRNAs on this specific pathway, the increased expression of MALAT1 and low levels of LINC00520 have been found to regulate the expression of EGFR receptor [107,118], while PICSAR targeted DUSP6, a negative regulator of ERK2 [113], and the downregulation of TINCR enhanced the ERK1/2-SP3 pathway [117]. Aside from RAS signaling, the Hippo-YAP pathway could represent another central signaling route in cSCC development, as the functional loss of the tumor suppressor FAT1 and the increased expression of lncRNA LINC01048 has been shown to favor the transcriptional activation of YAP1, promoting cell proliferation and survival [42,114].

Modulating the expression of the aforementioned molecular markers in various combinations could inhibit cell proliferation and survival, which may lead to the discovery of novel, efficient and reliable therapeutic approaches for cSCC. Pairing targeted therapy with conventional treatments may also represent a reliable strategy. At the moment, EGFR inhibitors in combination with radiation and chemotherapy have proved efficient against metastatic cSCC [8,128,129], while the functional restoration of TINCR, in combination with 5-aminolevulinic acid and photodynamic therapy triggered cell apoptosis and autophagy [117].

8. Conclusions

Cutaneous SCC is one of the most common types of neoplasia in the world, with a growing incidence every year. Due to its high mutational burden caused by cumulative UV light exposure, the identification and validation of specific key driver genes in cSCC has been difficult, however, commonly mutated genes have been found and established as reliable markers for this type of skin cancer. The search is still ongoing for novel markers that could stand as therapeutic targets, with microRNAs and lncRNAs at the forefront of recent studies. In the past few years, new therapeutic agents for cSCC have been developed, with EGFR and immune checkpoint inhibitors showing promising results. Moreover, these novel therapeutic approaches could partner with current treatment options (chemotherapy, radiation), giving clinicians the opportunity to adjust the treatment for high-risk patients. Unfortunately, despite the progress made in identifying specific reliable disease biomarkers and developing novel therapeutic approaches, cSCC continues to be lethal, if diagnosed in the advanced stages. Thus, the elucidation of the molecular mechanisms involved in the pathogenesis and evolution of this type of cancer represents a principal research objective at the moment, as it could lead to the identification of novel therapeutic targets, and to the improvement of patients' diagnosis and treatment.

Author Contributions: Conceptualization, S.D. and M.C.; Writing—original draft preparation, A.D.L.; Writing—review and editing, A.D.L., S.D. and M.C.; Figures and tables preparation—A.D.L.; S.D.; Supervision, M.C. All authors have read and agreed to the published version of the manuscript.

Funding: This research was funded by the Romanian Ministry of Research and Innovation, CCDI-UEFISCDI, grant number PN-III-P1-1.2-PCCDI-2017-0341/PATHDERM, within PNCDI III.

Conflicts of Interest: The authors declare no conflict of interest. The funders had no role in the design of the study; in the collection, analyses, or interpretation of data; in the writing of the manuscript, or in the decision to publish the results.

References

1. Apalla, Z.; Nashan, D.; Weller, R.B.; Castellsagué, X. Skin cancer: Epidemiology, disease burden, pathophysiology, diagnosis, and therapeutic approaches. *Dermatol. Ther.* **2017**, *7*, 5–19. [CrossRef]
2. Greinert, R. Skin cancer: New markers for better prevention. *Pathobiology* **2009**, *76*, 64–81. [CrossRef] [PubMed]
3. Global Burden of Disease Cancer Collaboration. Global, regional and national cancer incidence, mortality, years of life lost, years lived with disability, and disability-adjusted life-years for 29 cancer groups, 1990 to 2017. *JAMA Oncol.* **2019**, *5*, 1749–1768. [CrossRef] [PubMed]
4. Gloster, H.M.; Neal, K. Skin cancer in skin of color. *J. Am. Acad. Dermatol.* **2006**, *55*, 741–760. [CrossRef] [PubMed]
5. Lopez, A.T.; Liu, L.; Geskin, L. Molecular Mechanisms and Biomarkers of Skin Photocarcinogenesis. In *Human Skin Cancers: Pathways, Mechanisms, Targets and Treatments*; Blumenberg, M., Ed.; IntechOpen: London, UK, 2017; pp. 175–200.
6. Padilla, R.S.; Sebastian, S.; Jiang, Z.; Nindl, I.; Larson, R. Gene expression patterns of normal human skin, actinic keratosis, and squamous cell carcinoma: A spectrum of disease progression. *Arch. Dermatol.* **2010**, *146*, 288–293. [CrossRef]
7. Brougham, N.D.; Dennett, E.R.; Cameron, R.; Tan, S.T. The incidence of metastasis from cutaneous squamous cell carcinoma and the impact of its risk factors. *J. Surg. Oncol.* **2012**, *106*, 811–815. [CrossRef]
8. Palyca, P.; Koshenkov, V.P.; Mehnert, J.M. Developments in the treatment of locally advanced and metastatic squamous cell carcinoma of the skin: A rising unmet need. *Am. Soc. Clin. Oncol. Educ. Book* **2014**, *2014*, e397–e404. [CrossRef]
9. Durinck, S.; Ho, C.; Wang, N.J.; Liao, W.; Jakkula, L.R.; Collisson, E.A.; Pons, J.; Chan, S.W.; Lam, E.T.; Chu, C.; et al. Temporal dissection of tumorigenesis in primary cancers. *Cancer Discov.* **2011**, *1*, 137–143. [CrossRef]
10. Ashton, K.J.; Weinstein, S.R.; Maguire, D.J.; Griffiths, L.R. Chromosomal aberrations in squamous cell carcinoma and solar keratoses revealed by comparative genomic hybridization. *Arch. Dermatol.* **2003**, *139*, 876–882. [CrossRef]
11. Ratushny, V.; Gober, M.D.; Hick, R.; Ridky, T.W.; Seykora, J.T. From keratinocyte to cancer: The pathogenesis and modeling of cutaneous squamous cell carcinoma. *J. Clin. Investig.* **2012**, *122*, 464–472. [CrossRef]
12. Parekh, V.; Seykora, J.T. Cutaneous Squamous Cell Carcinoma. *Clin. Lab. Med.* **2017**, *37*, 503–525. [CrossRef] [PubMed]
13. García-Sancha, N.; Corchado-Cobos, R.; Pérez-Losada, J.; Cañueto, J. MicroRNA Dysregulation in Cutaneous Squamous Cell Carcinoma. *Int. J. Mol. Sci.* **2019**, *20*, 2181. [CrossRef]
14. Wang, Y.; Sun, B.; Wen, X.; Hao, D.; Du, D.; He, G.; Jiang, X. The Roles of lncRNA in Cutaneous Squamous Cell Carcinoma. *Front. Oncol.* **2020**, *10*, 158. [CrossRef] [PubMed]
15. Gordon, R. Skin cancer: An overview of epidemiology and risk factors. *Semin. Oncol. Nurs.* **2013**, *29*, 160–169. [CrossRef] [PubMed]

16. Pesch, B.; Ranft, U.; Jakubis, P.; Nieuwenhuijsen, M.J.; Hergemoller, A.; Unfried, K.; Jakubis, M.; Miskovic, P.; Keegan, T. Environmental arsenic exposure from a coal-burning power plant as a potential risk factor for nonmelanoma skin carcinoma: Results from a case-control study in the district of Prievidza, Slovakia. *Am. J. Epidemiol.* **2002**, *155*, 798–809. [CrossRef]
17. Nindl, I.; Gottschling, M.; Stockfleth, E. Human papillomaviruses and non-melanoma skin cancer: Basic virology and clinical manifestations. *Dis. Markers* **2007**, *23*, 247–259. [CrossRef]
18. Brenner, M.; Hearing, V.J. The protective role of melanin against UV damage in human skin. *Photochem. Photobiol.* **2008**, *84*, 539–549. [CrossRef]
19. Hussein, M.R. Ultraviolet radiation and skin cancer: Molecular mechanisms. *J. Cutan. Pathol.* **2005**, *32*, 191–205. [CrossRef]
20. Beani, J.C. Ultraviolet A-induced DNA damage: Role in skin cancer. *Bull. Acad. Natl. Med.* **2014**, *198*, 273–295.
21. Nilsen, L.T.; Hannevik, M.; Veierod, M.B. Ultraviolet exposure from indoor tanning devices: A systematic review. *BJD* **2016**, *174*, 730–740. [CrossRef]
22. Wehner, M.R.; Shive, M.L.; Chren, M.M.; Han, J.; Qureshi, A.A.; Linos, E. Indoor tanning and non-melanoma skin cancer: Systematic review and meta-analysis. *BMJ* **2012**, *345*, e5909. [CrossRef] [PubMed]
23. Breitbart, E.W.; Greinert, R.; Volkmer, B. Effectiveness of information campaigns. *Prog. Biophys. Mol. Biol.* **2006**, *92*, 167–172. [CrossRef] [PubMed]
24. Green, A.; Williams, G.; Neale, R.; Hart, V.; Leslie, D.; Parsons, P.; Marks, G.C.; Gaffney, P.; Battistutta, D.; Frost, C.; et al. Daily sunscreen application and betacarotene supplementation in prevention of basal-cell and squamous-cell carcinomas of the skin: A randomized controlled trial. *Lancet* **1999**, *354*, 723–729. [CrossRef]
25. Green, A.C.; Williams, G.M.; Logan, V.; Strutton, G.M. Reduced melanoma after regular sunscreen use: Randomized trial follow-up. *J. Clin. Oncol.* **2011**, *29*, 257–263. [CrossRef]
26. Gandini, S.; Sera, F.; Cattaruzza, M.S.; Pasquini, P.; Zanetti, R.; Masini, C.; Boyle, P.; Melchi, C.F. Meta-analysis of risk factors for cutaneous melanoma: III. Family history, actinic damage and phenotypic factors. *Eur. J. Cancer* **2005**, *41*, 2040–2059. [CrossRef]
27. South, A.P.; Purdie, K.J.; Watt, S.A.; Haldenby, S.; den Breems, N.; Dimon, M.; Arron, S.T.; Kluk, M.J.; Aster, J.C.; McHugh, A.; et al. NOTCH1 mutations occur early during cutaneous squamous cell carcinogenesis. *J. Investig. Dermatol.* **2014**, *134*, 2630–2638. [CrossRef]
28. Pickering, C.R.; Zhou, J.H.; Lee, J.J.; Drummond, J.A.; Peng, S.A.; Saade, R.E.; Tsai, K.Y.; Curry, J.L.; Tetzlaff, M.T.; Lai, S.Y.; et al. Mutational landscape of aggressive cutaneous squamous cell carcinoma. *Clin. Cancer Res. J. Am. Assoc. Cancer Res.* **2014**, *20*, 6582–6592. [CrossRef]
29. Li, Y.Y.; Hanna, G.J.; Laga, A.C.; Haddad, R.I.; Lorch, J.H.; Hammerman, P.S. Genomic analysis of metastatic cutaneous squamous cell carcinoma. *Clin. Cancer Res. J. Am. Assoc. Cancer Res.* **2015**, *21*, 1447–1456. [CrossRef]
30. Yilmaz, A.S.; Ozer, H.G.; Gillespie, J.L.; Allain, D.C.; Bernhardt, M.N.; Furlan, K.C.; Castro, L.T.F.; Peteres, S.B.; Nagarajan, P.; Kang, S.Y.; et al. Differential mutation frequencies in metastatic cutaneous squamous cell carcinomas versus primary tumors. *Cancer* **2017**, *123*, 1184–1193. [CrossRef]
31. Inman, G.J.; Wang, J.; Nagano, A.; Alexandrov, L.; Purdie, K.J.; Taylor, R.G.; Sherwood, V.; Thomson, J.; Hogan, S.; Spender, L.C.; et al. The genomic landscape of cutaneous SCC reveals drivers and a novel azathioprine associated mutational signature. *Nat. Commun.* **2018**, *9*, 3667. [CrossRef]
32. Kim, E.J.; Park, J.S.; Um, S.J. Identification and characterization of HIPK2 interacting with p73 and modulating functions of the p53 family in vivo. *J. Biol. Chem.* **2002**, *277*, 32020–32028. [CrossRef] [PubMed]
33. Lee, C.S.; Bhaduri, A.; Mah, A.; Johnson, W.L.; Ungewickell, A.; Aros, C.J.; Nguyen, C.B.; Rios, E.J.; Siprashvili, Z.; Straight, A.; et al. Recurrent point mutations in the kinetochore gene KNSTRN in cutaneous squamous cell carcinoma. *Nat. Genet.* **2014**, *46*, 1060–1062. [CrossRef] [PubMed]
34. Brown, V.L.; Harwood, C.A.; Crook, T.; Cronin, J.G.; Kelsell, D.P.; Proby, C.M. p16INK4a and p14ARF tumor suppressor genes are commonly inactivated in cutaneous squamous cell carcinoma. *J. Investig. Dermatol.* **2004**, *122*, 1284–1292. [CrossRef] [PubMed]
35. Nagarajan, P.; Ivan, D. Cutaneous squamous cell carcinomas: Focus on high-risk features and molecular alterations. *Glob Derm.* **2016**, *3*, 359–365.
36. Kern, F.; Niault, T.; Baccarini, M. Ras and Raf pathways in epidermis development and carcinogenesis. *Br. J. Cancer* **2011**, *104*, 229–234. [CrossRef]

37. Toll, A.; Salgado, R.; Yébenes, M.; Martín-Ezquerra, G.; Gilaberte, M.; Baró, T.; Solé, F.; Alameda, F.; Espinet, B.; Pujol, R.M. Epidermal growth factor receptor gene numerical aberrations are frequent events in actinic keratoses and invasive cutaneous squamous cell carcinomas. *Exp. Dermatol.* **2010**, *19*, 151–153. [CrossRef]
38. Su, F.; Viros, A.; Milagre, C.; Trunzer, K.; Bollag, G.; Spleiss, O.; Reis-Filho, J.S.; Kong, X.; Koya, R.C.; Flaherty, K.T.; et al. RAS mutations in cutaneous squamous-cell carcinomas in patients treated with BRAF inhibitors. *N. Engl. J. Med.* **2012**, *366*, 207–215. [CrossRef]
39. Ashford, B.G.; Clark, J.; Gupta, R.; Iyer, N.G.; Yu, B.; Ranson, M. Reviewing the genetic alterations in high-risk cutaneous squamous cell carcinoma: A search for prognostic markers and therapeutic targets. *Head Neck* **2017**, *39*, 1462–1469. [CrossRef]
40. Moriyama, M.; Durham, A.D.; Moriyama, H.; Hasegawa, K.; Nishikawa, S.I.; Radtke, F.; Osawa, M. Multiple roles of Notch signaling in the regulation of epidermal development. *Dev. Cell* **2008**, *14*, 594–604. [CrossRef]
41. Okuyama, R.; Tagami, H.; Aiba, S. Notch signaling: Its role in epidermal homeostasis and in the pathogenesis of skin diseases. *J. Dermatol. Sci.* **2008**, *49*, 187–194. [CrossRef]
42. Santos-de-Frutos, K.; Segrelles, C.; Lorz, C. Hippo Pathway and YAP Signaling Alterations in Squamous Cancer of the Head and Neck. *J. Clin. Med.* **2019**, *8*, 2131. [CrossRef] [PubMed]
43. Harwood, C.A.; Proby, C.M.; Inman, G.J.; Leigh, I.M. The promise of genomics and the development of targeted therapies for cutaneous squamous cell carcinoma. *Acta Derm. Venereol.* **2016**, *96*, 3–16. [CrossRef]
44. Lawrence, M.S.; Stojanov, P.; Mermel, C.H.; Robinson, J.T.; Garraway, L.A.; Golub, T.R.; Meyerson, M.; Gabriel, S.B.; Lander, E.S.; Getz, G. Discovery and saturation analysis of cancer genes across 21 tumour types. *Nature* **2014**, *505*, 495–501. [CrossRef] [PubMed]
45. Maertens, O.; Cichowski, K. An expanding role for RAS GTPase activating proteins (RAS GAPs) in cancer. *Adv. Biol. Regul.* **2014**, *55*, 1–14. [CrossRef] [PubMed]
46. Davoli, T.; Xu, A.W.; Mengwasser, K.E.; Sack, L.M.; Yoon, J.C.; Park, P.J.; Elledge, S.J. Cumulative haploinsufficiency and triplosensitivity drive aneuploidy patterns and shape the cancer genome. *Cell* **2013**, *155*, 948–962. [CrossRef] [PubMed]
47. Holland, P.; Willis, C.; Kanaly, S.; Glaccum, M.; Warren, A.; Charrier, K.; Murison, J.; Derry, J.; Virca, G.; Bird, T.; et al. RIP4 is an ankyrin repeat-containing kinase essential for keratinocyte differentiation. *Curr. Biol.* **2002**, *12*, 1424–1428. [CrossRef]
48. Stransky, N.; Egloff, A.M.; Tward, A.D.; Kostic, A.D.; Cibulskis, K.; Sivachenko, A.; Kryukov, G.V.; Lawrence, M.S.; Sougnez, C.; McKenna, A.; et al. The mutational landscape of head and neck squamous cell carcinoma. *Science* **2011**, *333*, 1157–1160. [CrossRef]
49. Kalay, E.; Sezgin, O.; Chellappa, V.; Mutlu, M.; Morsy, H.; Kayserili, H.; Kreiger, E.; Cansu, A.; Toraman, B.; Abdalla, E.M.; et al. Mutations in RIPK4 cause the autosomal-recessive form of popliteal pterygium syndrome. *Am. J. Hum. Genet.* **2012**, *90*, 76–85. [CrossRef]
50. Lee, P.; Jiang, S.; Li, Y.; Yue, J.; Gou, X.; Chen, S.Y.; Zhao, Y.; Schober, M.; Tan, M.; Wu, X. Phosphorylation of Pkp1 by RIPK 4 regulates epidermal differentiation and skin tumorigenesis. *EMBO J.* **2017**, *36*, 1963–1980. [CrossRef]
51. Gui, Y.; Guo, G.; Huang, Y.; Hu, X.; Tang, A.; Gao, S.; Wu, R.; Chen, C.; Li, X.; Zhou, L.; et al. Frequent mutations of chromatin remodeling genes in transitional cell carcinoma of the bladder. *Nat. Genet.* **2011**, *43*, 875–878. [CrossRef]
52. Ellis, M.J.; Ding, L.; Shen, D.; Luo, J.; Suman, V.J.; Wallis, J.W.; Van Tine, B.A.; Hoog, J.; Goiffon, R.J.; Goldstein, T.C.; et al. Whole-genome analysis informs breast cancer response to aromatase inhibition. *Nature* **2012**, *486*, 353–360. [CrossRef]
53. Je, E.M.; Lee, S.H.; Yoo, N.J.; Lee, S.H. Mutational and expressional analysis of MLL genes in gastric and colorectal cancers with microsatellite instability. *Neoplasma* **2013**, *60*, 188–195. [CrossRef] [PubMed]
54. Lee, J.J.; Sholl, L.M.; Lindeman, N.I.; Lee, J.J.; Sholl, L.M.; Lindeman, N.I.; Granter, S.R.; Laga, A.C.; Shivdasani, P.; Chin, G.; et al. Targeted next-generation sequencing reveals high frequency of mutations in epigenetic regulators across treatment-naïve patient melanomas. *Clin. Epigenet.* **2015**, *7*, 59. [CrossRef]
55. Gao, Y.B.; Chen, Z.L.; Li, J.G.; Hu, X.D.; Shi, X.J.; Sun, Z.M.; Zhang, F.; Zhao, Z.R.; Li, Z.T.; Liu, Z.Y.; et al. Genetic landscape of esophageal squamous cell carcinoma. *Nat. Genet.* **2014**, *46*, 1097–1102. [CrossRef]
56. Chung, C.H.; Guthrie, V.B.; Masica, D.L.; Tokheim, C.; Kang, H.; Richmon, J.; Agrawal, N.; Fakhry, C.; Quon, H.; Subramaniam, R.M.; et al. Genomic alterations in head and neck squamous cell carcinoma determined by cancer gene-targeted sequencing. *Ann. Oncol.* **2015**, *26*, 1216–1223. [CrossRef] [PubMed]

57. Peacock, J.W.; Takeuchi, A.; Hayashi, N.; Liu, L.; Tam, K.J.; Nakouzi, N.A.; Khazamipour, N.; Tombe, T.; Dejima, T.; Lee,, K.C.K.; et al. SEMA3C drives cancer growth by transactivating multiple receptor tyrosine kinases via Plexin B1. *EMBO Mol. Med.* **2018**, *10*, 219–238. [CrossRef]
58. Hui, D.H.F.; Tam, K.J.; Jiao, I.Z.F.; Ong, C.J. Semaphorin 3C as a Therapeutic Target in Prostate and Other Cancers. *Int. J. Mol. Sci.* **2019**, *20*, 774. [CrossRef] [PubMed]
59. Gomes, I.M.; Maia, C.J.; Santos, C.R. STEAP proteins: From structure to applications in cancer therapy. *Mol. Cancer Res.* **2012**, *10*, 573–587. [CrossRef]
60. Xue, X.; Bredell, B.X.; Anderson, E.R.; Martin, A.; Mays, C.; Nagao-Kitamoto, H.; Huang, S.; Győrffy, B.; Greenson, J.K.; Hardiman, K.; et al. Quantitative proteomics identifies STEAP4 as a critical regulator of mitochondrial dysfunction linking inflammation and colon cancer. *Proc. Natl. Acad. Sci. USA* **2017**, *114*, E9608–E9617. [CrossRef] [PubMed]
61. Briso, E.M.; Guinea-Viniegra, J.; Bakiri, L.; Rogon, Z.; Petzelbauer, P.; Eils, R.; Wolf, R.; Rincón, M.; Angel, P.; Wagner, E.F. Inflammation-mediated skin tumorigenesis induced by epidermal c-Fos. *Genes Dev.* **2013**, *27*, 1959–1973. [CrossRef] [PubMed]
62. Justilien, V.; Regala, R.P.; Tseng, I.C.; Walsh, M.P.; Batra, J.; Radisky, E.S.; Murray, N.R.; Fields, A.P. Matrix metalloproteinase-10 is required for lung cancer stem cell maintenance, tumor initiation and metastatic potential. *PLoS ONE* **2012**, *7*, e35040. [CrossRef] [PubMed]
63. Zhu, Z.; Di, J.; Lu, Z.; Gao, K.; Zheng, J. Rap2BGTPase: Structure, functions, and regulation. *Tumor Biol. J. Int. Soc. Oncodev. Biol. Med.* **2016**, *37*, 7085–7093. [CrossRef] [PubMed]
64. Le Duff, M.; Gouju, J.; Jonchere, B.; Guillon, J.; Toutain, B.; Boissard, A.; Henry, C.; Guette, C.; Lelievre, E.; Coqueret, O. Regulation of senescence escape by the cdk4-EZH2-AP2M1 pathway in response to chemotherapy. *Cell Death Dis.* **2018**, *9*, 1–15. [CrossRef]
65. Croce, C.M.; Calin, G.A. miRNAs, cancer, and stem cell division. *Cell* **2005**, *122*, 6–7. [CrossRef] [PubMed]
66. Gregory, R.I.; Shiekhattar, R. MicroRNA biogenesis and cancer. *Cancer Res.* **2005**, *65*, 3509–3512. [CrossRef]
67. Ponting, C.P.; Oliver, P.L.; Reik, W. Evolution and functions of long noncoding RNAs. *Cell* **2009**, *136*, 629–641. [CrossRef] [PubMed]
68. Fang, Y.; Fullwood, M.J. Roles, Functions, and Mechanisms of Long Non-coding RNAs in Cancer. *Genom. Proteom. Bioinf.* **2016**, *14*, 42–54. [CrossRef] [PubMed]
69. Yamane, K.; Jinnin, M.; Etoh, T.; Kobayashi, Y.; Shimozono, N.; Fukushima, S.; Masuguchi, S.; Maruo, K.; Inoue, Y.; Ishihara, T.; et al. Down-regulation of miR-124/-214 in cutaneous squamous cell carcinoma mediates abnormal cell proliferation via the induction of ERK. *J. Mol. Med.* **2013**, *91*, 69–81. [CrossRef]
70. Roskoski, R., Jr. Targeting ERK1/2 protein-serine/threonine kinases in human cancers. *Pharmacol. Res.* **2019**, *142*, 151–168. [CrossRef]
71. Suiqing, C.; Min, Z.; Lirong, C. Overexpression of phosphorylated-STAT3 correlated with the invasion and metastasis of cutaneous squamous cell carcinoma. *J. Dermatol.* **2005**, *32*, 354–360. [CrossRef]
72. Toll, A.; Salgado, R.; Espinet, B.; Diaz-Lagares, A.; Hernandez-Ruiz, E.; Andrades, E.; Sandoval, J.; Esteller, M.; Pujol, R.M.; Hernandez-Munoz, I. MiR-204 silencing in intraepithelial to invasive cutaneous squamous cell carcinoma progression. *Mol. Cancer* **2016**, *15*, 53. [CrossRef] [PubMed]
73. Gastaldi, C.; Bertero, T.; Xu, N.; Bourget-Ponzio, I.; Lebrigand, K.; Fourre, S.; Popa, A.; Cardot-Leccia, N.; Meneguzzi, G.; Sonkoly, E.; et al. miR-193b/365a cluster controls progression of epidermal squamous cell carcinoma. *Carcinogenesis* **2014**, *35*, 1110–1120. [CrossRef] [PubMed]
74. Neu, J.; Dziunycz, P.J.; Dzung, A.; Lefort, K.; Falke, M.; Denzler, R.; Freiberger, S.N.; Iotzova-Weiss, G.; Kuzmanov, A.; Levesque, M.P.; et al. miR-181a decelerates proliferation in cutaneous squamous cell carcinoma by targeting the proto-oncogene KRAS. *PLoS ONE* **2017**, *12*, e0185028. [CrossRef] [PubMed]
75. Luo, Q.; Li, W.; Zhao, T.; Tian, X.; Liu, Y.; Zhang, X. Role of miR-148a in cutaneous squamous cell carcinoma by repression of MAPK pathway. *Arch. Biochem. Biophys.* **2015**, *583*, 47–54. [CrossRef] [PubMed]
76. Bertero, T.; Bourget-Ponzio, I.; Puissant, A.; Loubat, A.; Mari, B.; Meneguzzi, G.; Auberger, P.; Barbry, P.; Ponzio, G.; Rezzonico, R. Tumor suppressor function of miR-483-3p on squamous cell carcinomas due to its pro-apoptotic properties. *Cell Cycle* **2013**, *12*, 2183–2193. [CrossRef]
77. Fleming, J.L.; Gable, D.L.; Samadzadeh-Tarighat, S.; Cheng, L.; Yu, L.; Gillespie, J.L.; Toland, A.E. Differential expression of miR-1, a putative tumor suppressing microRNA, in cancer resistant and cancer susceptible mice. *PeerJ* **2013**, *1*, e68. [CrossRef]

78. Yu, X.; Li, Z. The role of miRNAs in cutaneous squamous cell carcinoma. *J. Cell. Mol. Med.* **2016**, *20*, 3–9. [CrossRef]
79. Lefort, K.; Brooks, Y.; Ostano, P.; Cario-Andre, M.; Calpini, V.; Guinea-Viniegra, J.; Albinger-Hegyi, A.; Hoetzenecker, W.; Kolfschoten, I.; Wagner, E.F.; et al. A miR-34a-SIRT6 axis in the squamous cell differentiation network. *EMBO J.* **2013**, *32*, 2248–2263. [CrossRef]
80. Li, S.; Luo, C.; Zhou, J.; Zhang, Y. MicroRNA-34a directly targets high-mobility group box 1 and inhibits the cancer cell proliferation, migration and invasion in cutaneous squamous cell carcinoma. *Exp. Ther. Med.* **2017**, *14*, 5611–5618. [CrossRef]
81. Jia, G.; Su, L.; Singhal, S.; Liu, X. Emerging roles of SIRT6 on telomere maintenance, DNA repair, metabolism and mammalian aging. *Mol. Cell. Biochem.* **2012**, *364*, 345–350. [CrossRef]
82. Bouchie, A. First microRNA mimic enters clinic. *Nat. Biotechnol.* **2013**, *31*, 577. [CrossRef] [PubMed]
83. Beg, M.S.; Brenner, A.J.; Sachdev, J.; Borad, M.; Kang, Y.K.; Stoudemire, J.; Smith, S.; Bader, A.G.; Kim, S.; Hong, D.S. Phase I study of MRX34, a liposomal miR-34a mimic, administered twice weekly in patients with advanced solid tumors. *Investig. New Drugs* **2017**, *35*, 180–188. [CrossRef]
84. Folini, M.; Gandellini, P.; Longoni, N.; Profumo, V.; Callari, M.; Pennati, M.; Colecchia, M.; Supino, R.; Veneroni, S.; Salvioni, R.; et al. miR-21: An oncomir on strike in prostate cancer. *Mol. Cancer* **2010**, *9*, 12. [CrossRef] [PubMed]
85. Lohcharoenkal, W.; Harada, M.; Loven, J.; Meisgen, F.; Landen, N.X.; Zhang, L.; Lapins, J.; Mahapatra, K.D.; Shi, H.; Nissinen, L.; et al. MicroRNA-203 Inversely Correlates with Differentiation Grade, Targets c-MYC, and Functions as a Tumor Suppressor in cSCC. *J. Investig. Dermatol.* **2016**, *136*, 2485–2494. [CrossRef] [PubMed]
86. Zhou, J.; Liu, R.; Luo, C.; Zhou, X.; Xia, K.; Chen, X.; Zhou, M.; Zou, Q.; Cao, P.; Cao, K. MiR-20a inhibits cutaneous squamous cell carcinoma metastasis and proliferation by directly targeting LIMK1. *Cancer Biol. Ther.* **2014**, *15*, 1340–1349. [CrossRef] [PubMed]
87. Wang, S.H.; Zhou, J.D.; He, Q.Y.; Yin, Z.Q.; Cao, K.; Luo, C.Q. MiR-199a inhibits the ability of proliferation and migration by regulating CD44-Ezrin signaling in cutaneous squamous cell carcinoma cells. *Int. J. Clin. Exp. Pathol.* **2014**, *7*, 7131–7141. [PubMed]
88. Xu, N.; Zhang, L.; Meisgen, F.; Harada, M.; Heilborn, J.; Homey, B.; Grander, D.; Stahle, M.; Sonkoly, E.; Pivarcsi, A. MicroRNA-125b down-regulates matrix metallopeptidase 13 and inhibits cutaneous squamous cell carcinoma cell proliferation, migration, and invasion. *J. Biol. Chem.* **2012**, *287*, 29899–29908. [CrossRef]
89. Martin, T.A.; Harrison, G.; Mansel, R.E.; Jiang, W.G. The role of the CD44/ezrin complex in cancer metastasis. *Crit. Rev. Oncol. Hematol.* **2003**, *46*, 165–186. [CrossRef]
90. Chen, C.; Zhao, S.; Karnad, A.; Freeman, J.W. The biology and role of CD44 in cancer progression: Therapeutic implications. *J. Hematol. Oncol.* **2018**, *11*, 64. [CrossRef]
91. Konicke, K.; Lopez-Luna, A.; Munoz-Carrillo, J.L.; Servin-Gonzalez, L.S.; Flores-de la Torre, A.; Olasz, E.; Lazarova, Z. The microRNA landscape of cutaneous squamous cell carcinoma. *Drug Discov. Today* **2018**, *23*, 864–870. [CrossRef]
92. Gong, Z.; Zhou, F.; Shi, C.; Xiang, T.; Zhou, C.K.; Wang, Q.Q.; Jiang, Y.S.; Gao, S.F. miRNA-221 promotes cutaneous squamous cell carcinoma progression by targeting PTEN. *Cell. Mol. Biol. Lett.* **2019**, *24*, 9. [CrossRef] [PubMed]
93. Li, X.; Huang, K.; Yu, J. Inhibition of microRNA-21 upregulates the expression of programmed cell death 4 and phosphatase tensin homologue in the A431 squamous cell carcinoma cell line. *Oncol. Lett.* **2014**, *8*, 203–207. [CrossRef]
94. Darido, C.; Georgy, S.R.; Wilanowski, T.; Dworkin, S.; Auden, A.; Zhao, Q.; Rank, G.; Srivastava, S.; Finlay, M.J.; Papenfuss, A.T.; et al. Targeting of the tumor suppressor GRHL3 by a miR-21-dependent proto-oncogenic network results in PTEN loss and tumorigenesis. *Cancer Cell* **2011**, *20*, 635–648. [CrossRef] [PubMed]
95. Olasz, E.B.; Seline, L.N.; Schock, A.M.; Duncan, N.E.; Lopez, A.; Lazar, J.; Flister, M.J.; Lu, Y.; Liu, P.; Sokumbi, O.; et al. MicroRNA-135b Regulates Leucine Zipper Tumor Suppressor 1 in Cutaneous Squamous Cell Carcinoma. *PLoS ONE* **2015**, *10*, e0125412. [CrossRef] [PubMed]
96. Vecchione, A.; Baldassarre, G.; Ishii, H.; Nicoloso, M.S.; Belletti, B.; Petrocca, F.; Zanesi, N.; Fong, L.Y.; Battista, S.; Guarnieri, D.; et al. Fez1/Lzts1 absence impairs Cdk1/Cdc25C interaction during mitosis and predisposes mice to cancer development. *Cancer Cell* **2007**, *11*, 275–289. [CrossRef]

97. Chen, B.; Pan, W.; Lin, X.; Hu, Z.; Jin, Y.; Chen, H.; Ma, G.; Qiu, Y.; Chang, L.; Hua, C.; et al. MicroRNA-346 functions as an oncogene in cutaneous squamous cell carcinoma. *Tumor Biol.* **2015**, *37*, 2765–2771. [CrossRef]
98. Bruegger, C.; Kempf, W.; Spoerri, I.; Arnold, A.W.; Itin, P.H.; Burger, B. MicroRNA expression differs in cutaneous squamous cell carcinomas and healthy skin of immunocompetent individuals. *Exp. Dermatol.* **2013**, *22*, 426–428. [CrossRef]
99. Yu, J.; Peng, H.; Ruan, Q.; Fatima, A.; Getsios, S.; Lavker, R.M. MicroRNA-205 promotes keratinocyte migration via the lipid phosphatase SHIP2. *FASEB J.* **2010**, *24*, 3950–3959. [CrossRef]
100. Tian, J.; Shen, R.; Yan, Y.; Deng, L. miR-186 promotes tumor growth in cutaneous squamous cell carcinoma by inhibiting apoptotic protease activating factor-1. *Exp. Ther. Med.* **2018**, *16*, 4010–4018. [CrossRef]
101. Shakeri, R.; Kheirollahi, A.; Davoodi, J. Apaf-1: Regulation and function in cell death. *Biochimie* **2017**, *135*, 111–125. [CrossRef]
102. Wang, A.; Landen, N.X.; Meisgen, F.; Lohcharoenkal, W.; Stahle, M.; Sonkoly, E.; Pivarcsi, A. MicroRNA-31 is overexpressed in cutaneous squamous cell carcinoma and regulates cell motility and colony formation ability of tumor cells. *PLoS ONE* **2014**, *9*, e103206. [CrossRef] [PubMed]
103. Lin, N.; Zhou, Y.; Lian, X.; Tu, Y. MicroRNA-31 functions as an oncogenic microRNA in cutaneous squamous cell carcinoma cells by targeting RhoTBT1. *Oncol. Lett.* **2017**, *13*, 1078–1082. [CrossRef] [PubMed]
104. Zhou, L.; Wang, Y.; Zhou, M.; Zhang, Y.; Wang, P.; Li, X.; Yang, J.; Wang, H.; Ding, Z. HOXA9 inhibits HIF-1-mediated glycolysis through interacting with CRIP2 to repress cutaneous squamous cell carcinoma development. *Nat. Commun.* **2018**, *9*, 1480. [CrossRef] [PubMed]
105. Dinescu, S.; Ignat, S.; Lazar, A.D.; Constantin, C.; Neagu, M.; Costache, M. Epitranscriptomic Signatures in lncRNAs and Their Possible Roles in Cancer. *Genes* **2019**, *10*, 52. [CrossRef] [PubMed]
106. Zhou, L.; Zhu, Y.; Sun, D.; Zhang, Q. Emerging Roles of Long non-coding RNAs in The Tumor Microenvironment. *Int. J. Biol. Sci.* **2020**, *16*, 2094–2103. [CrossRef]
107. Zhang, Y.; Gao, L.; Ma, S.; Ma, J.; Wang, Y.; Li, S.; Hu, X.; Han, S.; Zhou, M.; Zhou, L.; et al. MALAT1-KTN1-EGFR regulatory axis promotes the development of cutaneous squamous cell carcinoma. *Cell Death Differ.* **2019**, *26*, 2061–2073. [CrossRef] [PubMed]
108. Li, S.S.; Zhou, L.; Gao, L.; Wang, Y.H.; Ding, Z.H. Role of long noncoding RNA MALAT1 promotes the occurrence and progression of cutaneous squamous cell carcinoma. *J. South. Med. Univ.* **2018**, *38*, 421–427.
109. Yu, G.-J.; Sun, Y.; Zhang, D.-W.; Zhang, P. Long non-coding RNA HOTAIR functions as a competitive endogenous RNA to regulate PRAF2 expression by sponging miR-326 in cutaneous squamous cell carcinoma. *Cancer Cell Int.* **2019**, *19*, 270. [CrossRef]
110. Sun, G.; Wang, Y.; Zhang, J.; Lin, N.; You, Y. MiR-15b/HOTAIR/p53 form a regulatory loop that affects the growth of glioma cells. *J. Cell. Biochem.* **2018**, *119*, 4540–4547. [CrossRef]
111. Hosono, Y.; Niknafs, Y.S.; Prensner, J.R.; Iyer, M.K.; Dhanasekaran, S.M.; Mehra, R.; Pitchiaya, S.; Tien, J.; Escara-Wilke, J.; Poliakov, A.; et al. Oncogenic role of THOR, a conserved cancer/testis long non-coding RNA. *Cell* **2017**, *171*, 1559. [CrossRef]
112. Liu, Z.; Wu, G.; Lin, C.; Guo, H.; Xu, J.; Zhao, T. IGF2BP1 over-expression in skin squamous cell carcinoma cells is essential for cell growth. *Biochem. Biophys. Res. Commun.* **2018**, *501*, 731–738. [CrossRef] [PubMed]
113. Piipponen, M.; Heino, J.; Kahari, V.-M.; Nissinen, L. Long non-coding RNA PICSAR decreases adhesion and promotes migration of squamous carcinoma cells by downregulating alpha 2 beta 1 and alpha 5 beta 1 integrin expression. *Biol. Open* **2018**, *7*, bio037044. [CrossRef] [PubMed]
114. Chen, L.; Chen, Q.; Kuang, S.; Zhao, C.; Yang, L.; Zhang, Y.; Zhu, H.; Yang, R. USF1-induced upregulation of LINC01048 promotes cell proliferation and apoptosis in cutaneous squamous cell carcinoma by binding to TAF15 to transcriptionally activate YAP1. *Cell Death Dis.* **2019**, *10*, 296. [CrossRef] [PubMed]
115. Li, F.; Liao, J.; Duan, X.; He, Y.; Liao, Y. Upregulation of LINC00319 indicates a poor prognosis and promotes cell proliferation and invasion in cutaneous squamous cell carcinoma. *J. Cell. Biochem.* **2018**, *119*, 10393–10405. [CrossRef] [PubMed]
116. Kretz, M.; Siprashvili, Z.; Chu, C.; Webster, D.E.; Zehnder, A.; Qu, K.; Lee, C.S.; Flockhart, R.J.; Groff, A.F.; Chow, J.; et al. Control of somatic tissue differentiation by the long non-coding RNA TINCR. *Nature* **2013**, *493*, 231–245. [CrossRef]
117. Zhou, W.; Zhang, S.; Li, J.; Li, Z.; Wang, Y.; Li, X. lncRNA TINCR participates in ALA-PDT-induced apoptosis and autophagy in cutaneous squamous cell carcinoma. *J. Cell. Biochem.* **2019**, *120*, 13893–13902. [CrossRef]

118. Mei, X.-L.; Zhong, S. Long noncoding RNA LINC00520 prevents the progression of cutaneous squamous cell carcinoma through the inactivation of the PI3K/Akt signaling pathway by downregulating EGFR. *Chin. Med. J.* **2019**, *132*, 454–465. [CrossRef]
119. Qiao, H.-P.; Gao, W.-S.; Huo, J.-X.; Yang, Z.-S. Long non-coding RNA GAS5 functions as a tumor suppressor in renal cell carcinoma. *Asian Pac. J. Cancer Prev.* **2013**, *14*, 1077–1082. [CrossRef]
120. Zhang, Z.; Zhu, Z.; Watabe, K.; Zhang, X.; Bai, C.; Xu, M.; Wu, F.; Mo, Y.-Y. Negative regulation of lncRNA GAS5 by miR-21. *Cell Death Differ.* **2013**, *20*, 1558–1568. [CrossRef]
121. Wang, T.-H.; Chan, C.-W.; Fang, J.-Y.; Shih, Y.-M.; Liu, Y.-W.; Wang, T.-C.V.; Chen, C.-Y. 2-O-Methylmagnolol upregulates the long non-coding RNA, GAS5, and enhances apoptosis in skin cancer cells. *Cell Death Dis.* **2017**, *8*, e2638. [CrossRef]
122. Rundhaug, J.E.; Fischer, S.M. Molecular mechanisms of mouse skin tumor promotion. *Cancers* **2010**, *2*, 436–482. [CrossRef] [PubMed]
123. Lewis, C.M.; Glisson, B.S.; Feng, L.; Wan, F.; Tang, X.; Wistuba, I.I.; El-Naggar, A.K.; Rosenthal, D.I.; Chambers, M.S.; Lustig, R.A.; et al. A phase II study of gefitinib for aggressive cutaneous squamous cell carcinoma of the head and neck. *Clin. Cancer Res.* **2012**, *18*, 1435–1446. [CrossRef] [PubMed]
124. Preneau, S.; Rio, E.; Brocard, A.; Peuvrel, L.; Nguyen, J.M.; Quéreux, G.; Dreno, B. Efficacy of cetuximab in the treatment of squamous cell carcinoma. *J. Dermatol. Treat.* **2014**, *25*, 424–427. [CrossRef] [PubMed]
125. Reigneau, M.; Robert, C.; Routier, E.; Mamelle, G.; Moya-Plana, A.; Tomasic, G.; Mateus, C. Efficacy of neoadjuvant cetuximab alone or with platinum salt for the treatment of unresectable advanced nonmetastatic cutaneous squamous cell carcinomas. *Br. J. Dermatol.* **2015**, *173*, 527–534. [CrossRef] [PubMed]
126. Kawakami, Y.; Nakamura, K.; Nishibu, A.; Yanagihori, H.; Kimura, H.; Yamamoto, T. Regression of cutaneous squamous cell carcinoma in a patient with chronic myeloid leukaemia on imatinib mesylate treatment. *Acta Derm. Venereol.* **2008**, *88*, 185–186. [CrossRef]
127. William, W.N., Jr.; Feng, L.; Ferrarotto, R.; Ginsberg, L.; Kies, M.; Lippman, S.; Glisson, B.; Kim, E.S. Gefitinib for patients with incurable cutaneous squamous cell carcinoma: A single-arm phase II clinical trial. *J. Am. Acad. Dermatol.* **2017**, *77*, 1110–1113. [CrossRef]
128. Yanagi, T.; Kitamura, S.; Hata, H. Novel Therapeutic Targets in Cutaneous Squamous Cell Carcinoma. *Front. Oncol.* **2018**, *8*, 79. [CrossRef]
129. De Lima, P.O.; Joseph, S.; Panizza, B.; Simpson, F. Epidermal growth factor receptor's function in cutaneous squamous cell carcinoma and its role as a therapeutic target in the age of immunotherapies. *Curr. Treat. Options Oncol.* **2020**, *21*, 9. [CrossRef]
130. Patel, R.; Chang, A.L.S. Immune Checkpoint Inhibitors for Treating Advanced Cutaneous Squamous Cell Carcinoma. *Am. J. Clin. Dermatol.* **2019**, *20*, 477–482. [CrossRef]
131. US Food and Drug Administration. FDA Approves Cemiplimab-Rwlc for Metastatic or Locally Advanced Cutaneous Squamous Cell Carcinoma. FDA Website. Available online: https://www.fda.gov/drugs/drug-approvals-and-databases (accessed on 22 June 2020).
132. Love, W.E.; Bernhard, J.D.; Bordeaux, J.S. Topical imiquimod or fluorouracil therapy for basal and squamous cell carcinoma: A systematic review. *Arch. Dermatol.* **2009**, *145*, 1431–1438. [CrossRef]
133. Fayne, R.; Nanda, S.; Nichols, A.; Shen, J. Combination Topical Chemotherapy for the Treatment of an Invasive Cutaneous Squamous Cell Carcinoma. *J. Drugs Dermatol.* **2020**, *19*, 202–204. [CrossRef] [PubMed]

© 2020 by the authors. Licensee MDPI, Basel, Switzerland. This article is an open access article distributed under the terms and conditions of the Creative Commons Attribution (CC BY) license (http://creativecommons.org/licenses/by/4.0/).

Article

Influence of Serum Vitamin D Level in the Response of Actinic Keratosis to Photodynamic Therapy with Methylaminolevulinate

Ricardo Moreno [1,*], Laura Nájera [2], Marta Mascaraque [3], Ángeles Juarranz [3], Salvador González [4] and Yolanda Gilaberte [5]

1 Dermatology Service, Hospital Univ. del Henares, Coslada, 28822 Madrid, Spain
2 Pathology Service, Hospital Puerta de Hierro, Majadahonda, 28222 Madrid, Spain; lauranajerabotello@hotmail.com
3 Department of Cellular Biology, Universidad Autónoma de Madrid, 28049 Madrid, Spain; marta.mascaraque@uam.es (M.M.); angeles.juarranz@uam.es (A.J.)
4 Medicine and Medical Specialties Department, Instituto Ramón y Cajal de Investigación Sanitaria (IRYCIS), Universidad de Alcalá, 28034 Madrid, Spain; salvagonrod@gmail.com
5 Dermatology Service, Hospital Univ. Miguel Servet, 50009 Zaragoza, Spain; ygilaberte@gmail.com
* Correspondence: alonsodecelada@gmail.com; Tel.: +34-911-912-000

Received: 16 December 2019; Accepted: 25 January 2020; Published: 1 February 2020

Abstract: In mouse models of squamous cell carcinoma, pre-treatment with calcitriol prior to photodynamic therapy with aminolevulinic acid (ALA) enhances tumor cell death. We have evaluated the association between vitamin D status and the response of actinic keratoses to photodynamic therapy with methylaminolevulinate. Twenty-five patients with actinic keratoses on the head received one session of photodynamic therapy with methylaminolevulinate. Biopsies were taken at baseline and six weeks after treatment. Immuno-histochemical staining was performed for VDR, P53, Ki67 and β-catenin. Basal serum 25(OH)D levels were determined. Cases with a positive histological response to treatment had significantly higher serum 25(OH)D levels (26.96 (SD 7.49) ngr/mL) than those without response (18.60 (SE 7.49) ngr/mL) ($p = 0.05$). Patients with a complete clinical response displayed lower basal VDR expression (35.71% (SD 19.88)) than partial responders (62.78% (SD 16.735)), ($p = 0.002$). Our results support a relationship between vitamin D status and the response of actinic keratoses to photodynamic therapy with methylaminolevulinate.

Keywords: photochemotherapy; methylaminolevulinate; actinic keratosis; vitamin d; calcitriol; vitamin d receptor

1. Introduction

Vitamin D (VD) is a prohormone involved in a wide variety of functions in the organism, and has been related with several types of cancer [1]. It has several known effects on epidermal carcinogenesis [2]: it regulates keratinocyte proliferation promoting its differentiation [3], and it prevents UV-induced mutations [4], enhancing mutation repair.

In humans, vitamin D is obtained mainly through exposure to sunlight which, in the epidermis, promotes transformation of 7-dehydrocholesterol into cholecalciferol or previtamin D3. Secondarily, cholecalciferol is hydroxylated in the liver to become 25(OH)D or calcidiol, then further hydroxylated in the kidney into 1,25(OH)D or calcitriol, the biologically active form of vitamin D [1]. Calcitriol acts on its intracellular receptor (VDR), which is present in almost all cell types in humans, and its signaling exerts antiproliferative, antiangiogenic, pro-differentiating and antiapoptotic effects [5].

Actinic keratoses (AKs) are skin areas of keratinocytic dysplasia representing a preoplastic state—or according to some authors, an in situ form—of cutaneous squamous cell carcinoma (SCC).

In AK, the severity of keratinocytic dysplasia is classified, as in other intraepidermal carcinomas (CIN for cervical, VIN for vulvar, AIN for anal intraepithelial neoplasia) into KIN (keratinocytic intraepidermal neoplasia) grade 1, 2 or 3 according to the presence of dysplastic keratinocytes in one third, two thirds or the complete thickness of the epidermis [6].

Photodynamic therapy (PDT) with aminolevulinic acid (ALA) or methyl-aminolevulinate (MAL) is effective in clearing keratinocytic dysplasia and reversing some of the molecular features of AK, such as the expression of mutant P53 [7]. In this therapy, AKs are treated with mentioned photosensitizers and exposed to specific wavelength light sources. ALA and MAL are precursors of protoporphyrin IX (PpIX), a molecule that selectively accumulates in dysplastic keratinocytes. Irradiation induces photobleaching of PpIX which is responsible for tumor cell death.

In SCC murine models, pre-treatment with topical vitamin D prior to ALA-PDT has been shown to enhance PpIX accumulation and tumor cell death [8]. This has also been observed in other rodent models with oral [9] or intraperitoneal [10] administration of calcitriol. In humans, the clinical response of AK to PDT, in a split-scalp trial comparing MAL-PDT alone vs. MAL-PDT with a pre-treatment of 15 days with topical calcipotriol (a synthetic derivative of calcitriol marketed to treat psoriasis), improved in the pretreated group [11]. Galimberti also demonstrated superior efficacy of daylight mediated MAL-PDT after pre-treatment with calcipotriol ointment [12].

We intended to explore if VD or its receptor play a role in the response of AK to PDT. Therefore, we designed a study to evaluate the association between the serum 25(OH)D level and the skin expression of VD receptor (VDR) in AK with the response to MAL-PDT at clinical, histological and immuno-histochemical levels.

2. Patients and Methods

2.1. Design

A prospective observational pilot study was designed to establish whether serum 25(OH)D level influences the response of AK to MAL-PDT in patients.

2.2. Ethics

The study was approved by the local Ethical Committee at Hospital Universitario Puerta de Hierro in Madrid (Spain). The written informed consent was obtained from all the subjects before being recruited for the study.

2.3. Subjects

Twenty-five patients were enrolled in the study. Inclusion criteria were as follows: having five or more neighboring AKs susceptible to be treated with MAL-PDT, located on the face or the scalp. Exclusion criteria were: unstable health conditions, such as cancer or immunosuppression; medical contraindications for the treatment, such as pregnancy or photosensitivity; allergy to the MAL or any of the excipients of the cream; and having received any treatment for face or scalp AKs within the last six months. Patients were recruited from March 2014 to September 2016. Variables such as age, gender, body mass index (BMI) and Fitzpatrick phototype were collected for each patient.

2.4. Treatment Protocol

Patients were treated with conventional MAL-PDT as follows. AKs were prepared for the treatment by removal of hyperkerosis through gentle use of sandpaper or curettage. Then, a 1 mm layer of MAL 160 mg/g cream (Metvix®, Galderma, Paris, France) was applied over each AK, spreading the remaining cream on the surrounding cancerization area. After incubation under occlusion for three hours, the whole area was exposed to a red LED device emitting at 630 nm (Aktilite®; PhotoCure, Oslo, Norway) with a fluence of 37 J/cm^2. After the treatment, patients were instructed to avoid sun exposure using the same SPF 50 sunscreen cream (Heliocare Airgel, IFC, Spain) when outdoors until the end of the study.

2.5. Clinical Evaluation

Clinical evaluation was assessed using digital photographs before and six weeks after the treatment. Clinical lesion response was evaluated by two dermatologists who measured the reduction in the AK number and the Olsen grade in the treated area. Patient response was classified in three categories: complete response, defined by a 75% to 100% reduction in number and improvement in the grade of Olsen of the AKs; partial response when the overall improvement in number and grade was lower than 75% and higher than 25%; and improvement was lower than 25% percent.

2.6. Biochemical Variables

Two blood tests were performed on each patient, the first one on the day of the PDT treatment and the second one 6 weeks later, at the end of the follow-up. Serum levels of 25(OH)D (ng/mL) (electrochemiluminescence, Roche Diagnostics, Madrid, Spain) were determined in the Central Laboratory of Madrid. VD deficiency was defined as serum 25(OH)D of 20 ng/mL or less, VD insufficiency as values of 20–30 ng/mL, and sufficiency over it [13].

2.7. Histological and Immune-Histochemical Variables

A 3 mm punch biopsy of the index lesion (the most severe AK in the area) was performed 1 week before the treatment and 6 weeks after it. The second biopsy was taken at a minimal distance from the first biopsy scar.

The skin samples were fixed (10% formalin), embedded in paraffin, sectioned (3 μm thickness) and stained by haematoxylin and eosin, and then simultaneously subjected to immunohistochemistry using the corresponding antibodies for detection of Ki67 (prediluted; Ventana Medical Systems Tucson, AZ, USA), vitamin D receptor (VDR), P53, and β-catenin antibodies (Cell Signaling Technologies, Leiden, The Nederlands). Representative sections were examined using positive and negative controls. Immuno-histochemical evaluation of P53, Ki-67, β-catenin and VDR was performed by identifying, in each section, the area with the highest levels of immunoexpression ("hot spots") and estimating the percentage of cells with nuclear positivity in a high-power field (×400). Intensity of VDR staining was semi-quantitatively assessed by classifying expression intensity into 4 categories: 0, absence of staining; 1, mild staining (0%–33% tumoral cell staining); 2, moderate staining (>33%–66%) and 3, intense staining (>66%–100% tumoral cell staining).

Histological and immuno-histochemical variables (histological diagnosis, histological subtype of AK, KIN grade as defined by Röwert-Huber et al. [6], β-catenin, P53, Ki67, VDR expressions and VDR intensity were evaluated by a pathologist, blind to the identity of the samples. Histological response of the AK index, assessed on hematoxylin-eosin stained sections, was defined as positive when complete clearance or at least a decrease in two KIN grades was achieved, and negative when absence of histological response or decrease of only one KIN grade was shown.

2.8. Statistical Analysis

Quantitative variables are expressed as mean and standard deviation (SD) and dichotomous variables as proportions. Associations between qualitative variables were assessed using Pearson's Chi-squared test or Fisher's exact test. Mann–Whitney U-test or Student's t-test for paired data was used to evaluate associations between quantitative variables. Pearson correlation coefficient was calculated to evaluate the linear correlation between two variables. Statistical significance was set at $p \leq 0.05$. Analyses were performed using SPSS Statistics (version 19.0: IBM, Armonk, NY, USA).

3. Results

3.1. Demographic Characteristics of the Sample

All twenty-five patients completed the study. However, one case was excluded from the histological analysis since the post-treatment biopsy revealed a collision of an actinic and a seborrheic keratosis. The mean age was 70.1 years (range 61–81) and 76% were males with Fitzpatrick phototype 3 (60%) or phototype 2 (40%). Most of the treated AKs were located on the scalp (64%) and 36% on the facial area (Table 1). The mean basal 25(OH)D serum levels were 25.37 (SD 9.86) ng/mL.

The severity of keratinocytic dysplasia was considered KIN3 in 7 lesions (29.17%), KIN2 in 10 (41.66%) and KIN1 in 7 (29.17%) AK.

Table 1. Sociodemographic and biochemical variables of the sample. (SD: standard deviation; BMI: body mass index.).

Variables (N = 25)		Frecuency	Mean (Range or SD)
Age (years)			70.1 (61–81)
Gender	Male	19/25 (76%)	
	Female	6/25 (24%)	
Phototype	II	10/25 (40%)	
	III	15/25 (60%)	
B.M.I. (kg/m^2)			30.1 (23.30–42.40)
Location of treated AK	Face	9/25 (36%)	
	Scalp	16/25 (64%)	
Serum 25(OH)D$_3$ (ng/mL)			25.37 (SD 9.86)

3.2. Clinical and Histological Response to PDT Per Lesion

As expected, PDT induced a significant reduction in the mean number of AKs per patient, from 7.80 (SD 2.79) to 2.8 (SD 1.61) ($p = 0.005$) (Figure 1). Overall clinical response was complete in 16 patients (64%) and partial in 9 (36%); there were no cases without response.

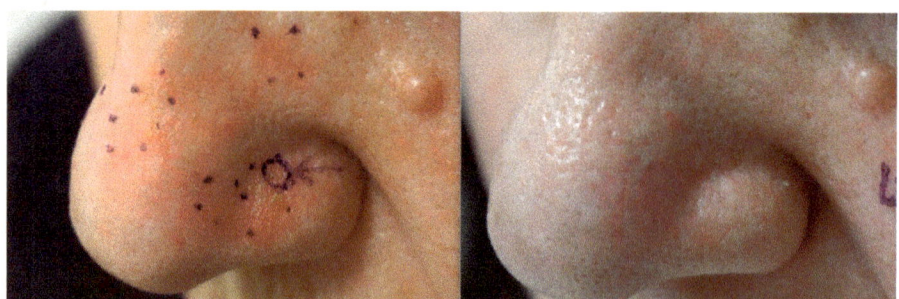

Figure 1. Complete clinical response to photodynamic therapy (PDT), as clearance of actinic keratoses in the nasal area of a patient six weeks after treatment.

Histological response was positive in 17 AK (70.8%) and negative in 7 AK (29.2%). Index AK exhibited basal KIN grade 3 in 29.17%, KIN 2 in 41.66%, and KIN 1 in 29.17% of the samples, and after treatment KIN grade was 3 in 8.33%, KIN 2 in 12.50%, KIN 1 in 16.67% and KIN 0 in 62.50% of the lesions, showing a significant improvement of the KIN grade ($p = 0.004$) Considering the KIN grade as a quantitative variable, PDT induced a significant decrease in the mean KIN grade, from 1.88 (SD 0.85) to 0.67 (SD 1.01) ($p = 0.000$).

PDT also induced a significant decrease in the mean of the immunostaining of Ki67 (57.08 (SD 27.10) to 26.88 (SD 19.27), $p = 0.001$) and P53 expression (59.17 (SD 27.72) to 26.39 (SD 24.54), $p = 0.001$). VDR expression increased after PDT but the differences were not statistically significant (56.67 (SD 20.36) to 66.67 (SD 22.00), $p = 0.062$) (Figure 2). No relevant differences were found in the rest of the immunological markers after PDT (Table 2).

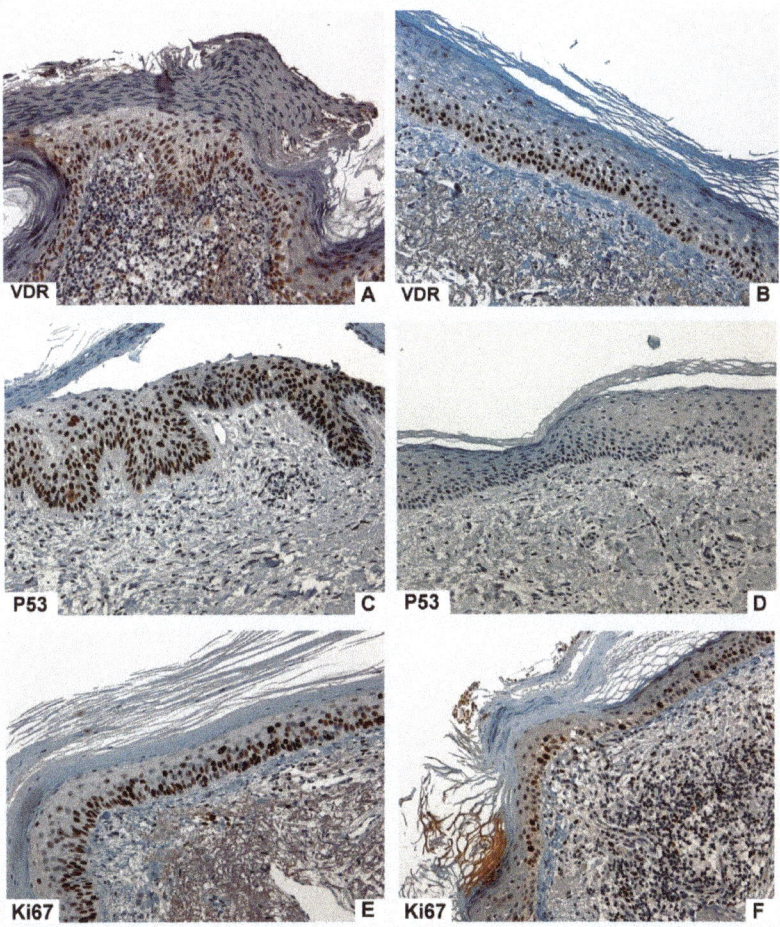

Figure 2. Actinic keratoses: immuno-histochemical response to MAL-PDT (methyl-aminolevulinate photodynamic therapy). Baseline vitamin D receptor (VDR) expression (**A**) did not significantly change after treatment (**B**). Baseline P53 (**C**) and Ki67 (**E**) expression significantly decreased (**D** and **F**, respectively) after PDT.

Table 2. Clinical, histological and immuno-histochemical variables of the sample, before and after MAL-PDT (methyl-aminolevulinate photodynamic therapy).

N = 24	Basal (mean, SD)	After PDT (mean, SD)	p
Clinical and histological variables			
AK number per patient	7.84 (SD 2.79)	2.80 (SD 1.61)	0.005
KIN grade (quantitative)	1.88 (0.85)	0.67 (1.01)	<0.001
KIN grade (qualitative)			
KIN 3	7 (29.17 %)	2 (8.33%)	0.004
KIN 2	10 (41.66%)	3 (12.50%)	
KIN 1	7 (29.17%)	4 (16.67%)	
KIN 0	0	15 (62.50%)	
Immunomarkers			
VDR expression (%)	56.67 (20.36)	66.67 (22.00)	0.062
VDR intensity (0–3)	1.96 (0.81)	2.08 (0.93)	0.479
β-catenin expression (%)	4.17 (5.69)	2.61 (4.59)	0.191
Ki67 expression (%)	57.08 (27.10)	26.88 (19.27)	0.000
P53 expression (%)	59.17 (27.72)	26.39 (24.54)	0.000

SD: Standard deviation; VDR: vitamin D receptor; KIN: keratinocytic intraepithelial neoplasia; AK: actinic keratosis.

3.3. Association of Variables with Overall Clinical Response

No statistically significant relationship was found between patient clinical response and age, gender, phototype, AK location and serum 25(OH)D. However, those patients with complete clinical response showed lower VDR expression (35.71, SD 19.88) than those with partial response (62.78, SD 16.74) ($p = 0.002$). Basal β-catenin, Ki67 and P53 expressions were not associated with the overall clinical outcome (Table 3).

3.4. Association of the Variables with the Histological Response

No statistically significant relationship was found between age, gender, phototype and location of the AK and the histological response. Patients who responded to PDT had significantly higher serum 25(OH)D levels (26.96 ng/mL, SD 7.49) than those without response (18.60, SD 7.49) ($p = 0.05$). Baseline expression of the explored immunomarkers was not associated with the histological response to PDT (Table 3).

Table 3. Influence of clinical and histological variables on overall clinical response of patients and histological response of AK to MAL-PDT.

	Patient Clinical Response			Histological Response		
	Partial Response (mean, SD) $n = 6$	Complete response (mean, SD) $n = 19$	p	Positive (mean, SD) $n = 17$	Negative (mean, SD) $n = 7$	p
Age (mean, SD)	71.47 (6.66)	69.67 (3.20)	0.53	69.88 (6.19)	73.86 (5.37)	0.153
Gender						
Male	14 (73.37%)	5 (26.30%)	1	14 (73.70%)	5 (26.30%)	0.608
Female	5 (83.30%)	1 (16.70%)		3 (60.00%)	2 (40.00%)	
Phototype						
II	6 (60.00%)	4 (40.00%)	0.175	5 (55.60%)	4 (44.4%)	0.356
III	13 (86.70%)	2 (13.30%)		12 (80.00%)	3 (20.00%)	
Location						
-Face	7 (77.80%)	2 (22.20%)	1	7 (87.50%)	1 (12.50%)	0.352
-Scalp	12 (75.00%)	4 (25.00%)		10 (62.50%)	6 (37.50%)	
Vitamin D (ng/ml)	24.42 (9.67)	27.67 (9.86)	0.483	26.96 (9.49)	18.60 (7.49)	0.05
VDR expression (%)	62.78 (16.74)	35.71 (19.88)	0.002	59.41 (18.53)	53.33 (25.03)	0.535
VDR intensity (0–3)	2.00 (0.77)	1.71 (0.95)	0.442	2.00 (0.79)	1.83 (0.98)	0.68
β-cat. expression (%)	4.39 (5.62)	3.86 (5.40)	0.832	5.53 (6.06)	17 (2.04)	0.103
Ki67 expression(%)	56.39 (28.12)	51.00 (32.20)	0.683	55.00 (29.42)	59.17 (22.00)	0.756
P53 expression (%)	64.41 (23.51)	59.00 (38.79)	0.701	64.69 (27.35)	52.00 (22.80)	0.361

4. Discussion

This study supports the relationship between 25(OH)D serum levels and the response of AK to MAL-PDT: VD deficient levels were found to be significantly associated to a lack of response in the reduction of the KIN grade of actinic keratoses, and patients whose AK exhibited a significantly lower VDR basal expression showed a complete clinical response to the treatment. Comparing the histological samples of AK in every patient before and after MAL-PDT, we observed a marginally significant increase in VDR expression after the treatment in addition to the already know reduction in P53 and Ki67 expression [7].

Thus, our findings suggest that a poorer response of AK to MAL-PDT is likely to be expected under a deficient VD status. The mechanisms by which VD may exert an effect on the response of AK to MAL-PDT are unknown. It has been demonstrated that VD promotes UV-induced mutation repair in keratinocytes through an up-regulation of functional P53 [14] and has several antitumoral effects on epidermal neoplasms through the immune system [15,16]. The transcriptional profile of healthy keratinocytes treated with 1,25(OH)D has been studied, showing the up-regulation of some 82 genes and down-regulation of 16 other genes; among those up-regulated were peptidilarginine deaminases, calicreins, serin-protease inhibitors, c-fos or Kruppel-like factor 4, all of which are involved in keratinocyte differentiation [17]. These findings illustrate a poorly understood pro-differentiation network over keratinocytes sustained by VD.

The increase on PpIX accumulation and consequent enhanced tumoral cell death induced by exposure to calcitriol or calcipotriol prior to ALA-PDT has been proposed as a possible mechanism in several previously published studies on murine models [8–10]. However, Bay et al. [18] exposed hairless mice to carcinogenic doses of UV-radiation, and found that pre-treatment with calcipotriol prior to MAL-PDT neither increased PpIX accumulation, as measured by fluorescence, nor delayed the onset of SCC compared to MAL-PDT without pre-treatment. Accordingly, VD ability to enhance AK response to PDT may not be related to an increased PpIX accumulation in keratinocytes, but to other mechanisms, perhaps involving its complex effects on keratinocyte proliferation and differentiation.

The relationship between 25(OH)D serum levels and VDR expression in keratinocytes has not been studied in humans. In swine models fed with a VD-deficient diet, VD insufficiency status resulted in a diffuse presence of VDR in the keratinocytic cytoplasm, whilst supplementation with VD up to serum levels of 25(OH)D sufficiency induced a preferentially nuclear location of the VDR [19]. Moreover, unpublished data of our group in several in vitro experiences with skin and vulvar squamous cell carcinoma cell lines (SCC-13 and A-431) reveal that these tumor cells exhibit a diffuse expression of

VDR in the cytoplasm that changes dramatically to a predominantly nuclear expression upon the addition of calcitriol to the cell cultures.

As for the discrepancy between the influence of variables on clinical and on histological responses of AK to PDT, it has already been proven that clinical and histological classifications of AK do not accurately match and that conclusions must not be drawn about the histology of AK lesions from their clinical appearance [20]. Hence, we can infer that parallel findings should not necessarily be expected between clinical and histological approaches.

According to our results comparing immunohystochemical features of our samples before and after the treatment, MAL-PDT might be able to increase VDR expression in the nucleus of the keratinocytes in AK. This may improve keratinocyte sensitivity to serum VD, thus providing an additional plausible way through which PDT and VD interact to generate a synergic antitumoral effect on keratinocytic neoplasms such as AK.

An important limitation of this is study is the small sample. A larger sample is warranted to strengthen the evidence provided by our findings.

In summary, our findings imply that serum VD could be considered a modulator of the response of AK to MAL-PDT. Determination of serum 25(OH)D might be appropriate in patients with AK eligible for treatment with MAL-PDT, in order to predict their intrinsic ability to respond to it, and to select those patients who could benefit from VD supplementation prior to treatment. More research is needed, firstly to confirm our results and secondly to establish if VD supplementation in deficient patients previous to PDT might improve its efficacy.

Author Contributions: Conceptualization, S.G., A.J. and Y.G.; Methodology, Y.G.; Validation, Y.G.; Formal Analysis, R.M. and Y.G.; Investigation, R.M., M.M., A.J. and L.N.; Resources, M.M., L.N., Á.J. and R.M.; Data Curation, R.M.; Writing—Original Draft Preparation, R.M; Writing—Review & Editing, Y.G. and S.G.; Project Administration, Á.J. and Y.G.; Funding Acquisition, Á.J., S.G. and Y.G. All authors have read and agreed to the published version of the manuscript.

Funding: This research has been partially supported by a grant from the Carlos III Health Institute, Ministry of Science, Innovation and University, Spain (PI18/00858), and by Galderma.

Acknowledgments: The authors want to acknowledge to Galderma for their economic funding and scientific support in the development of this research line, for their help with English editing of this article and for the Metvix cream employed in the study. Our work was also supported by Cantabria Labs who provided the Heliocare Airgel sunscreen for the subjects of the study.

Conflicts of Interest: The sponsors had no role in the design, execution, interpretation, or writing of the study.

References

1. Giovannucci, E. The epidemiology of vitamin D and cancer incidence and mortality: A review (United States). *Cancer Causes Control.* **2005**, *16*, 83–95. [CrossRef] [PubMed]
2. Kamradt, J.; Rafi, L.; Mitschele, T.; Meineke, V.; Gärtner, B.C.; Tilgen, W.; Holick, M.F.; Reichrath, J. Analysis of the Vitamin D system in Cutaneous Malignancies. *Recent Results Cancer Res.* **2003**, *164*, 259–269. [PubMed]
3. Hu, L.; Bikle, D.D.; Oda, Y. Reciprocal role of vitamin D receptor on β-catenin regulated keratinocyte proliferation and differentiation. *J. Steroid Biochem. Mol. Biol.* **2014**, *144*, 237–241. [CrossRef] [PubMed]
4. Wong, G.; Gupta, R.; Dixon, K.; Deo, S.; Choong, S.; Halliday, G.; Bishop, J.; Ishizuka, S.; Norman, A.; Posner, G.; et al. 1,25-Dihydroxyvitamin D and three low-calcemic analogs decrease UV-induced DNA damage via the rapid response pathway. *J. Steroid Biochem. Mol. Boil.* **2004**, *89*, 567–570. [CrossRef] [PubMed]
5. Vuolo, L.; Di Somma, C.; Faggiano, A.; Colao, A.A. Vitamin D and Cancer. *Front. Endocrinol.* **2012**, *23*, 3–58. [CrossRef] [PubMed]
6. Röwert-Huber, J.; Patel, M.J.; Forschner, T.; Ulrich, C.; Eberle, J.; Kerl, H.; Sterry, W.; Stockfleth, E. Actinic keratosis is an early in situ squamous cell carcinoma: A proposal for reclassification. *Br. J. Dermatol.* **2007**, *156*, 8–12. [CrossRef] [PubMed]
7. Bagazgoitia, L.; Santos, J.C.; Juarranz, A.; Jaén, P. Photodynamic therapy reduces the histological features of actinic damage and the expression of early oncogenic markers. *Br. J. Dermatol.* **2011**, *165*, 144–151. [CrossRef] [PubMed]

8. Rollakanti, K.; Anand, S.; Maytin, E.V. Topical calcitriol prior to photodynamic therapy enhances treatment efficacy in non-melanoma skin cancer mouse models. *Proc. SPIE Int. Soc. Opt. Eng.* **2015**, *9308*, 93080Q. [PubMed]
9. Anand, S.; Rollakanti, K.R.; Horst, R.L.; Hasan, T.; Maytin, E.V. Combination of oral vitamin D3 with photodynamic therapy enhances tumor cell death in a murine model of cutaneous squamous cell carcinoma. *Photochem. Photobiol.* **2014**, *90*, 1126–1135. [PubMed]
10. Anand, S.; Wilson, C.; Hasan, T.; Maytin, E.V. Vitamin D3 enhances the apoptotic response of epithelial tumors to aminolevulinate-based photodynamic therapy. *Cancer Res.* **2011**, *71*, 6040–6050. [CrossRef] [PubMed]
11. Torezan, L.; Grinblat, B.; Haedersdal, M.; Valente, N.; Festa-Neto, C.; Szeimies, R.M. A randomized split-scalp study comparing calcipotriol-assisted methyl aminolaevulinate photodynamic therapy (MAL-PDT) with conventional MAL-PDT for the treatment of actinic keratosis. *Br. J. Dermatol.* **2018**, *179*, 829–835. [CrossRef] [PubMed]
12. Galimberti, G.N. Calcipotriol as pretreatment prior to daylight-mediated photodynamic therapy in patients with actinic keratosis: A case series. *Photodiagnosis Photodyn. Ther.* **2018**, *21*, 172–175. [CrossRef] [PubMed]
13. Holick, M.F.; Binkley, N.C.; Bischoff-Ferrari, H.A.; Gordon, C.M.; Hanley, D.A.; Heaney, R.P.; Murad, M.H.; Weaver, C.M. Evaluation, Treatment, and Prevention of Vitamin D Deficiency: An Endocrine Society Clinical Practice Guideline. *J. Clin. Endocrinol. Metab.* **2011**, *96*, 1911–1930. [CrossRef] [PubMed]
14. Reichrath, J.; Reichrath, S.; Heyne, K.; Vogt, T.; Roemer, K. Tumor suppression in skin and other tissues via cross-talk between vitamin D- and p53-signaling. *Front. Physiol.* **2014**, *5*. [CrossRef] [PubMed]
15. Bikle, D.D. The vitamin D receptor: A tumor suppressor in skin. *Single Mol. Single Cell Seq.* **2014**, *810*, 282–302.
16. Bikle, D.D.; Jiang, Y. The Protective Role of Vitamin D Signaling in Non-Melanoma Skin Cancer. *Cancers* **2013**, *5*, 1426–1438. [CrossRef] [PubMed]
17. Lu, J.; Goldstein, K.M.; Chen, P.; Huang, S.; Gelbert, L.M.; Nagpal, S. Transcriptional Profiling of Keratinocytes Reveals a Vitamin D-Regulated Epidermal Differentiation Network. *J. Investig. Dermatol.* **2005**, *124*, 778–785. [CrossRef] [PubMed]
18. Bay, C.; Togsverd-Bo, K.; Lerche, C.M.; Haedersdal, M. Skin tumor development after UV irradiation and photodynamic therapy is unaffected by short-term pretreatment with 5-fluorouracil, imiquimod and calcipotriol. *An experimental hairless mouse study. J. Photochem. Photobiol. B: Boil.* **2016**, *154*, 34–39. [CrossRef] [PubMed]
19. Trowbridge, R.M.; Mitkov, M.V.; Hunter, W.J.; Agrawal, D.K. Vitamin D Receptor and CD86 Expression in the Skin of Vitamin D-Deficient Swine. *Exp. Mol. Pathol.* **2014**, *96*, 42–47. [CrossRef] [PubMed]
20. Schmitz, L.; Kahl, P.; Majores, M.; Bierhoff, E.; Stockfleth, E.; Dirschka, T. Actinic keratosis: Correlation between clinical and histological classification systems. *J. Eur. Acad. Dermatol. Venereol.* **2016**, *30*, 1303–1307. [CrossRef] [PubMed]

© 2020 by the authors. Licensee MDPI, Basel, Switzerland. This article is an open access article distributed under the terms and conditions of the Creative Commons Attribution (CC BY) license (http://creativecommons.org/licenses/by/4.0/).

Article

In Vivo Reflectance Confocal Microscopy-Diagnostic Criteria for Actinic Cheilitis and Squamous Cell Carcinoma of the Lip

Mihai Lupu [1], Ana Caruntu [2,3,*], Daniel Boda [1,4] and Constantin Caruntu [4,5]

1. Dermatology Research Laboratory, "Carol Davila" University of Medicine and Pharmacy, 050474 Bucharest, Romania; lupu.g.mihai@gmail.com (M.L.); daniel.boda@yahoo.com (D.B.)
2. Department of Oral and Maxillofacial Surgery, "Carol Davila" Central Military Emergency Hospital, 010825 Bucharest, Romania
3. Faculty of Medicine, "Titu Maiorescu" University, 031593 Bucharest, Romania
4. Department of Dermatology, "Prof. N.C. Paulescu" National Institute of Diabetes, Nutrition and Metabolic Diseases, 011233 Bucharest, Romania; costin.caruntu@gmail.com
5. Department of Physiology, "Carol Davila" University of Medicine and Pharmacy, 050474 Bucharest, Romania
* Correspondence: ana.caruntu@gmail.com; Tel.: +40-72-2345-344

Received: 22 May 2020; Accepted: 24 June 2020; Published: 25 June 2020

Abstract: Actinic cheilitis (AC) is one of the most frequent pathologies to affect the lips. Studies show that the most commonplace oral malignancy, squamous cell carcinoma (SCC), often emerges from AC lesions. Invasive diagnostic techniques performed on the lips carry a high risk of complications, but reflectance confocal microscopy (RCM), a non-invasive skin imaging technique, may change the current diagnostic pathway. This retrospective study was aimed at consolidating the RCM diagnostic criteria for AC and lip SCC. The study was conducted in two tertiary care centers in Bucharest, Romania. We included adults with histopathologically confirmed AC and SCC who also underwent RCM examination. Of the twelve lesions included in the study, four were AC and eight were SCC. An atypical honeycomb pattern and the presence of target cells in the epidermis were RCM features associated with AC. SCC was typified by the presence of complete disruption of the epidermal architecture and dermal inflammatory infiltrates. The mean blood vessel diameter in SCC was 18.55 µm larger than that in AC ($p = 0.006$) and there was no significant difference ($p = 0.64$) in blood vessel density, as measured by RCM, between SCC and AC. These data confirm that RCM can be useful for the *in vivo* distinction between AC and lip SCC.

Keywords: actinic cheilitis; squamous cell carcinoma; in vivo; reflectance confocal microscopy; lip neoplasms

1. Introduction

Lips constitute a special location for the development of numerous skin lesions due to their frequent exposure to exogenous factors, such as ultraviolet light, chemical, and biological agents.

Actinic cheilitis (AC) is one of the most often occurring pathologies that affects the lips [1]. Occupational activities not considered, studies report an AC prevalence between 0.2% and 0.45% [2,3].

Upon clinical examination, AC has a broad spectrum of presentation, comprised of pale, dry, scaly lips, chronic ulcerations and erosions [4], blurring of the vermillion-skin border [2,5–10], white [11,12] and red [2,5,9,10] areas, and vermillion atrophy [5–8]. Palpation reveals a fine sandpaper-like feeling [4], often accompanied by a stinging or burning and an inelastic or tight sensation of the lip [4,8].

AC differential diagnosis includes inflammatory disorders such as eczema, benign leukoplakia, lichen planus, granulomatous cheilitis, and xerosis with chronic irritation [13,14].

Some studies show that squamous cell carcinoma (SCC) represents approximately 90% of all oral malignancy cases [15] while others estimate that 95% of SCCs of the lips emerge from ACs [12,16].

Changes suggestive of AC progression to SCC of the lip include thickening and induration of keratotic AC patches, the appearance of nodules with rapid growth and/or ulceration associating bleeding and pain [4,8,17–20]. Lip squamous cell carcinoma is also much more prone to metastasis than cutaneous SCC (0.5–3% vs. 3–20%) [21–23].

A number of techniques hold promise for the early detection, aggressiveness profile and monitoring of keratinocyte carcinomas. The observed differences between normal, inflammatory and malignant keratinocyte proteomic profiles are likely to unearth novel markers for SCC, in terms of diagnosis and monitoring, and could maybe even come to the aid of targeted therapies [24–27].

Although the gold standard diagnostic technique for AC and lip SCC is the histopathological examination of a biopsy specimen, the anatomic characteristics of the lips increase the risk of postoperative bleeding and infection. Additionally, considering the cosmetic importance of this area, noninvasive diagnostic techniques are useful for selecting the biopsy site, thus avoiding repeated biopsies, and in some cases even acting as a surrogate for histopathology.

Imaging techniques, such as dermoscopy and *in vivo* reflectance confocal microscopy (RCM), continue to highlight diagnostic and prognostic criteria for AC and SCC [28,29]. The lips are an ideal site for RCM examination, due to a thinner epidermis when compared to other body sites. Because early detection and swift therapy remain the two most important factors influencing the long-term survival of these patients [30], we designed a retrospective study with the aim of consolidating previous observations regarding the RCM diagnostic criteria for AC and lip SCC.

2. Materials and Methods

2.1. Subjects

Patients with histopathologically confirmed lesions of actinic cheilitis or lip SCC were included in the study. Patients' records were retrieved from the electronic database of the Dermatology Research Laboratory, "Carol Davila" University of Medicine and Pharmacy, in Bucharest. The study was conducted in accordance with the Declaration of Helsinki, and the protocol was approved by the Local Ethics Committee (No. 25/27.11.2017). All participants gave written informed consent as part of their investigation and treatment procedures, at the time of their registration.

2.2. RCM Imaging and Analysis

Despite it being a retrospective study, the RCM imaging protocol was the same for all lesions, as it is a well-established protocol in this clinic when examining non-melanocytic lesions.

Confocal imaging was done with a commercially available confocal microscope (Vivascope® 1500, Caliber ID, Rochester, NY, USA) which uses a near-infrared laser diode with a maximum power output of 20 mW. The device and image acquisition protocol have been described elsewhere [31].

Vertical imaging via Vivastack® was performed by capturing a series of images of 0.5 × 0.5 mm with 3 μm increments, in depth. Horizontal mosaics (via Vivablock®) of 4 × 4 mm were captured at different depths of the lesions. Mapping started at the stratum corneum and was continued to the papillary dermis.

Due to the problematic separation of epidermal layers on RCM, we adopted the methodology of Peppelman et al. [32], so that the first appearance of nucleated cells, regardless of cell size and shape, was considered to be the stratum granulosum (SG). Since the SG is only a few cell layers thick, three steps in depth below this point was considered as the stratum spinosum (SS).

Diagnostic RCM criteria for AC and lip SCC were selected based on previously published data (Table 1) [33–42].

Table 1. Reflectance confocal microscopy (RCM) criteria for the diagnosis of actinic cheilitis (AC) and lip squamous cell carcinoma (SCC).

Epidermis	
Ulceration	Dark areas, with irregular and well-defined borders, filled with amorphous material and cellular debris.
Hyperkeratosis/scale	Increased thickness of the stratum corneum seen as areas of amorphous, variably refractive material, and reduced resolution of deeper structures.
Parakeratosis	Presence of individual polygonal, sharply delineated, nucleated cells in the stratum corneum.
Atypical honeycomb pattern SG/SS *	Cells with irregular shape and size showing bright cell borders, arranged in a distorted fashion, deviating from the normal honeycomb pattern.
Architectural disarrangement SG/SS*	Disarray of the normal architecture of the superficial skin layers with unevenly dispersed hyper-refractive granular particles and cells, in which the honeycomb or cobblestone patterns are no longer visible.
Targetoid cells SS/SG*	A large cell resembling a target, either with a bright center and dark peripheral halo or a dark center and a bright rim surrounded by a dark peripheral halo. The first one corresponds histologically to large dyskeratotic keratinocytes separated from adjacent cells by a retraction halo, and the second type to dyskeratotic keratinocytes containing a pyknotic nucleus.
Dendritic cells	Large cells with obvious dendrites connected to them.
Dermal-epidermal junction	
Increased vessel diameter	Blood vessel diameter larger than 5 µm.
Increased vessel density	More than 5 blood vessels per 0.5 × 0.5 mm RCM image.
Dermis	
Solar elastosis	Lace-like material adjacent to hyper-refractive, thickened collagen bundles.
Inflammatory cells	Hyper-refractive, small structures, of 8–10 µm in diameter.
Dendritic cells	Large cells with obvious dendrites connected to them.
Atypical keratinocytes (speckled/nucleated)	Round to polygonal cells with a dark nucleus and speckled appearance.
Nest-like structures	Defined, irregular, discohesive, aggregates of cells larger than inflammatory cells.
Keratin pearls	Whorl-shaped, hyper-refractive, speckled structures.

* SG/SS, stratum granulosum/stratum spinosum.

RCM images were then evaluated for these features by an RCM user (LM) with 4 years of experience with this technique. The observer systematically evaluated the lesions for the presence or absence of individual RCM criteria, but was not blinded to the histopathological diagnosis. Furthermore, the mean vessel diameter and blood vessel density per individual confocal image (500 × 500 µm) were determined for both AC and lip SCC lesions. Based on previously published data [32], an increased vascular diameter was defined as a diameter greater than 5 µm and an increased blood vessel density as more than 5 vessels per single confocal image (500 × 500 µm). Blood vessel diameter measurements were taken on scale calibrated images using the open source software package ImageJ. A line, perpendicular to the axis of the vessel, was drawn from side to side in the widest visible part of the blood vessel and the result was recorded. This measurement was repeated for every visible vessel in the chosen RCM image. For every case, these two parameters were measured in 3 different single confocal images at approximately the same depth for every lesion, and the highest value for each of these two parameters was recorded for the case.

2.3. Histopathology

All lesions included in this study were surgically excised after RCM investigation and histologically confirmed as either AC or lip SCC by an experienced pathologist on haematoxylin–eosin (H&E) stained paraffin sections. SCC was defined based on the presence of an invasive component.

2.4. Statistical Analysis

The analysis was conducted in order to assess how various observed RCM criteria were associated with either AC or SCC. It comprised descriptive statistics and Fisher's exact test to analyze the differences between subgroups. The differences in blood vessel diameter and vessel density per individual confocal image in AC and SCC lesions were measured using an independent t-test.

All data analyses were conducted using the statistical software package SPSS Inc. (v20, Chicago, IL, USA).

3. Results

Twelve subjects (10 males and 2 females) with a mean age of 66.82 ± 9.87 (range 43–80) years were included in the study.

In these patients, a total of 12 biopsy-proven lesions were evaluated with RCM, of which four ACs and eight invasive lip SCCs. The lesions were all located on the vermillion of the lower lip. In the actinic cheilitis subgroup, there were two smokers (one male and one female) and in the SCC subgroup there were three smokers (all male). None of the subjects included in this study were immunosuppressed.

The degree of differentiation for the SCC lesions included in the study, along with gender, age, immune and smoking status, are illustrated in Table 2.

Table 2. Degree of differentiation, immune and smoking status in the SCC subgroup.

Sex	Age	Smoking	Immune Status	SCC Degree of Differentiation
male	43	yes	immunocompetent	moderately differentiated
female	59	no	immunocompetent	well differentiated
male	80	no	immunocompetent	well differentiated
male	69	no	immunocompetent	moderately differentiated
male	71	yes	immunocompetent	well differentiated
male	66	no	immunocompetent	moderately differentiated
male	68	yes	immunocompetent	moderately differentiated
male	65	no	immunocompetent	moderately differentiated

SCC, squamous cell carcinoma.

3.1. RCM Features for Differentiating between AC and Lip SCC

Hyperkeratosis (4/4), parakeratosis (3/4), an atypical honeycomb pattern (4/4), and the presence of dyskeratotic, targetoid cells within the epidermis (4/4) were RCM features present in virtually all the examined AC lesions (Figure 1). Total epidermal disarray, dendritic and atypical cells in the dermis, and tumor nests in the dermis were found in none of the AC lesions.

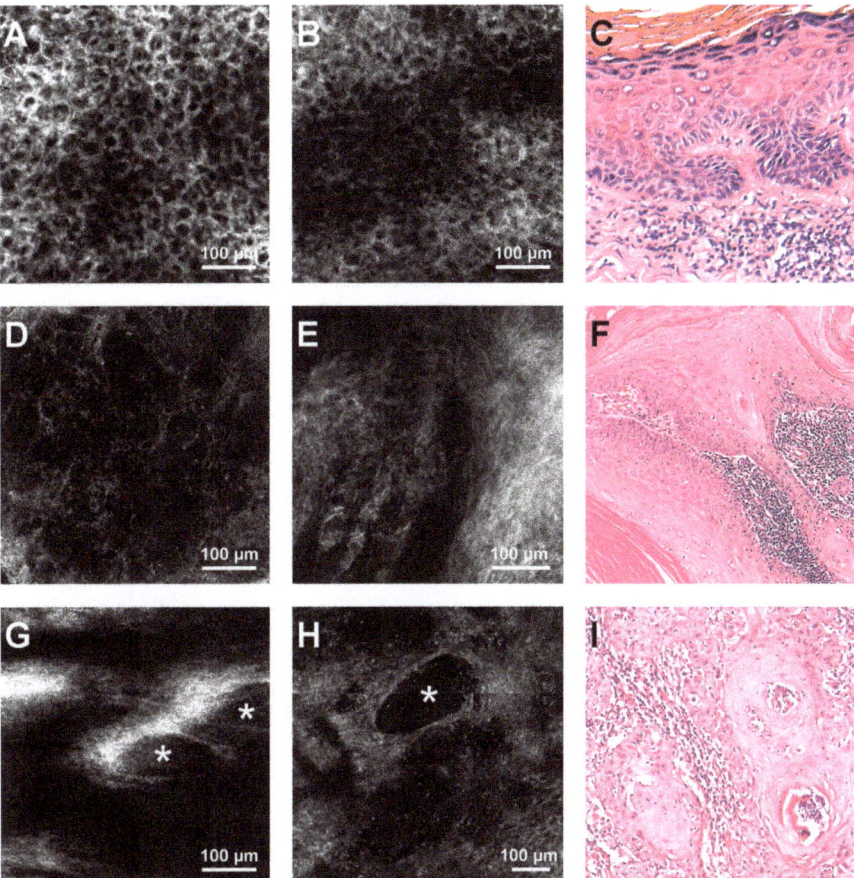

Figure 1. Representative reflectance confocal microscopy (RCM) images of actinic cheilitis (AC) and lip squamous cell carcinoma (SCC), and histological correspondents. (**A**) RCM image of an atypical honeycomb pattern of the stratum granulosum, seen in an AC lesion. (**B**) RCM image at the stratum spinosum showing an atypical honeycomb pattern, which can be seen in either AC or SCC. (**C**) Histopathology image illustrating parakeratosis, atypical keratinocytes in the stratum granulosum and spinosum, spongiosis, and intradermal inflammatory infiltrate in an AC lesion (haematoxylin-eosin, cropped, original magnification 40×). (**D**) RCM image of the complete architectural disarray in the granular layer of a lip SCC. (**E**) RCM image showing disarray in the stratum spinosum, in a SCC lesion. (**F**) Histopathological image displaying infiltrative atypical polygonal squamous cells with distinct cell borders, abundant eosinophilic cytoplasm, and large vesicular nuclei with moderate nuclear pleomorphism in a SCC (haematoxylin-eosin, cropped, original magnification 40×) (**G,H**). RCM images showing tumor nests (white asterisks) surrounded by white areas corresponding to fibrosis at the level of the dermis, in a SCC. (**I**) Histopathology image illustrating invasive SCC nests and strands of atypical polygonal squamous cells surrounded by intradermal inflammatory infiltrate in a lip SCC (haematoxylin–eosin, cropped, original magnification 100×).

Lip SCCs were typified in confocal examination by the presence of the complete disruption of the epidermal architecture (8/8) and the presence of dermal inflammatory infiltrate (6/8). Atypical cells in the dermis and dermal tumor nests could be detected upon RCM examination in only half of the SCC lesions (Figure 1). Dendritic cells, most probably corresponding to Langerhans cells, were seen in three out of eight SCCs. Although in SCCs with a pigmentary component, melanocytes can be seen in RCM, none of the tumors included in our study had a clinically or dermatoscopically visible pigmentary component. Even so, we cannot exclude the possibility that some of the observed dendritic cells could indeed be melanocytes.

Ulceration, hyperkeratosis/scale, and dermal solar elastosis were present in both AC and SCC with similar frequencies.

Table 3 shows the frequencies of the various RCM criteria in the studied AC and lip SCC lesions. Table 4 contains the frequencies of the observed RCM criteria according to SCC degree of differentiation. None of the RCM criteria varied significantly between well and moderately differentiated SCC, as assessed by the Chi-square test.

Table 3. Frequencies of RCM criteria for AC and lip SCC.

RCM Criteria, N (%)	Histopathological Diagnosis	
	AC (N = 4)	Lip SCC (N = 8)
Ulceration	3 (75%)	7 (87.5%)
Hyperkeratosis/scale	4 (100%)	7 (87.5%)
Parakeratosis	3 (75%)	3 (37.5%)
Atypical honeycomb pattern	4 (100%)	0 (0%)
Epidermal disarray	0 (0%)	8 (100%)
Target cells in the epidermis	4 (100%)	1 (12.5%)
Dendritic cells in the epidermis	1 (25%)	0 (0%)
Solar elastosis	2 (50%)	5 (62.5%)
Dermal inflammatory cells	2 (50%)	6 (75%)
Dendritic cells in the dermis	0 (0%)	3 (37.5%)
Atypical cells in the dermis	0 (0%)	4 (50%)
Tumor nests in the dermis	0 (0%)	4 (50%)

RCM, reflectance confocal microscopy; AC, actinic cheilitis; SCC, squamous cell carcinoma.

Table 4. Frequencies of RCM criteria according to SCC degree of differentiation.

RCM Criteria, N (%)	Squamous Cell Carcinoma		p
	Well Differentiated (N = 3)	Moderately Differentiated (N = 5)	
Ulceration	3 (100%)	4 (80%)	0.408
Hyperkeratosis/scale	3 (100%)	4 (80%)	0.408
Parakeratosis	0 (0%)	3 (60%)	0.09
Atypical honeycomb pattern	0 (0%)	0 (0%)	-
Epidermal disarray	3 (100%)	5 (100%)	-
Target cells in the epidermis	1 (33.3%)	0 (0%)	0.168
Dendritic cells in the epidermis	0 (0%)	0 (0%)	-
Solar elastosis	2 (66.7%)	3 (60%)	0.85
Dermal inflammatory cells	3 (100%)	3 (60%)	0.206
Dendritic cells in the dermis	0 (0%)	3 (60%)	0.09
Atypical cells in the dermis	2 (66.7%)	2 (40%)	0.465
Tumor nests in the dermis	2 (66.7%)	2 (40%)	0.465

RCM, reflectance confocal microscopy; SCC, squamous cell carcinoma.

The assessment of associations between RCM criteria and the diagnosis of AC or lip SCC revealed that an atypical honeycomb pattern ($p = 0.002$, Fisher's exact test) and the presence of target keratinocytes in the epidermis ($p = 0.01$, Fisher's exact test) were strongly associated with AC (Figure 2), while complete epidermal disarray ($p = 0.002$, Fisher's exact test) was characteristic for lip SCC. The RCM features for AC/SCC discrimination in this study have been summarized in Table 5.

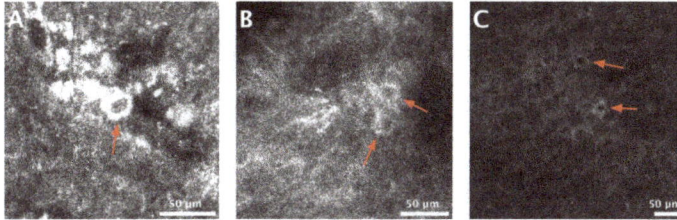

Figure 2. Target cells in the epidermis of actinic cheilitis (AC) lesions. (**A–C**) Reflectance confocal microscopy (RCM) images showing target cells (red arrows), corresponding to dyskeratotic keratinocytes, at the level of the epidermis in AC lesions.

Table 5. Specific RCM features associated with AC and lip SCC.

RCM Criteria	p
Actinic cheilitis	
Atypical honeycomb pattern	0.002
Target cells in the epidermis	0.01
Lip squamous cell carcinoma	
Complete epidermal disarray	0.002

AC, actinic cheilitis; SCC, squamous cell carcinoma; RCM, reflectance confocal microscopy.

3.2. Vascularization in AC and SCC Lesions

The vessel diameter and number of blood vessels per single RCM image were higher in SCC (Figure 3). The mean blood vessel diameter in SCCs was 18.55 μm larger than that in AC lesions ($p = 0.006$). There was no significant difference ($p = 0.64$) in blood vessel density, as measured by RCM, between SCC and AC lesions in our sample (Table 6).

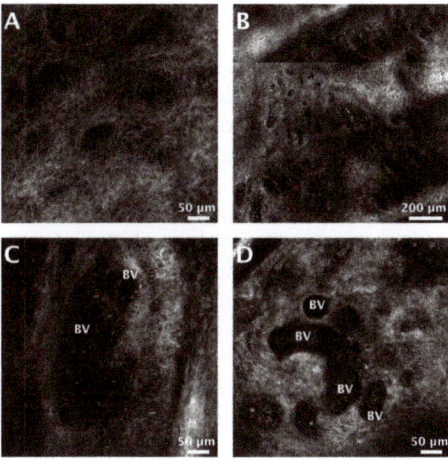

Figure 3. Blood vessel density and blood vessel dilation in actinic cheilitis (AC) and lip squamous cell carcinoma (SCC). (**A**) Reflectance confocal microscopy (RCM) image showing blood vessels (dark areas in the honeycomb) at the level of the dermal-epidermal junction (DEJ), in an AC lesion. (**B**) RCM mosaic (1305 × 1305 μm) displaying a high density of blood vessels at the DEJ in a SCC. (**C**) RCM image of dilated blood vessels (BV) filled with moderately-refractile particles (corresponding to blood cells) in an AC lesion. (**D**) RCM image showing markedly dilated blood vessels (BV) at dermal level in a SCC.

Table 6. Blood vessels characteristics of patients with AC and lip SCC.

	Histopathological Diagnosis		p
	AC	Lip SCC	
	Mean ± SD	Mean ± SD	
Mean blood vessel diameter (μm)	19.26 ± 5.67	37.81 ± 12.77	0.006
Mean number of blood vessels	8.25 ± 1.89	8.88 ± 2.53	0.64

AC, actinic cheilits; SCC, squamous cell carcinoma; SD, standard deviation.

4. Discussions

Aside from the particular functional and cosmetic significance of the lips, several conditions ranging from benign infections to dysplasias, and the potentially fatal SCC, may develop in this area. Hence, the examination of the vermillion and lip mucosa is an important part of the dermatological examination.

While there are numerous publications regarding RCM imaging of various skin conditions ranging from tumors to infections and inflammatory conditions [33–35,43–47], there are only a few reports in the literature related to the non-invasive diagnosis of lip lesions. Whilst RCM knowledge and experience in the field of non-melanoma skin cancer is constantly expanding [48–51], obtaining biopsies is still an invasive procedure with its own limitations, mainly due to cosmetic reasons and the risk of sampling errors.

An obvious limitation of our study is the small sample size of only 12 lesions. It remains difficult to have a large sample size of SCCs in non-invasive diagnostic studies. The hyperkeratotic scale makes the evaluation of SCC difficult with either dermoscopy or RCM, hence the limited advantage of these techniques in clinically evident SCCs. Additionally worth mentioning is the difficulty raised by the need for histopathological confirmation in this type of study. The most important struggle in this field remains in distinguishing between SCC in situ and invasive SCC in the case of clinically similar lesions.

Ulrich et al. [41] defined the RCM criteria for AC as: disruption of stratum corneum, hyperkeratosis, parakeratosis, atypical honeycomb pattern at the SS and SG, dermal solar elastosis, dilated blood vessels, and the presence of inflammatory cells in the upper dermis. In our study, the most common RCM feature for both AC and SCC was the presence of keratinocyte pleomorphism resulting in either an atypical honeycomb pattern in the case of AC or a complete disruption of the epidermal architecture in SCC. Furthermore, testing for the association between RCM criteria and these two entities, we uncovered that target cells are significantly associated with AC. These findings are in accordance with previous studies which examined the RCM appearance of actinic keratoses and cutaneous SCCs [32,38,52,53].

As opposed to previous studies [54], dermal dyskeratotic keratinocytes were observed in only 50% of SCC lesions included in our study, either as isolated, scattered cells or as tumor nests. However, similar to earlier research [54], we found dermal dendritic cells in under half of the lip SCCs (3/8) and these results do not allow for a significant association between this element and lip SCC. On the other hand, we observed the presence of inflammatory cells in the dermis in half of the ACs and almost all (6/8) of the SCCs, which is in line with other studies [54], and reveals traits of the tumor microenvironment [55]. While not statistically significant, Hartmann et al. [56] reported an increased frequency of RCM-observed dermal tumor nests and peritumoral inflammatory infiltrates in moderately differentiated SCCs compared to well differentiated SCCs. In our data, none of the RCM criteria were significantly different between well and moderately differentiated SCC. In our case, we attribute this to the small sample size.

When analyzing AC and lip SCC vasculature, we found a significantly increased mean vascular diameter and a larger blood vessel density for SCC compared to AC. Our results are in accordance with other studies, and can be explained by the high metabolic needs of a tumor, which determines vascular dilation and neovascularization [32,57,58].

To conclude, this study, building upon previous research, confirms several RCM criteria which can be used to distinguish between AC and lip SCC in vivo. This warrants further prospective, large sample-size studies, which will form the basis for the development of protocols for the correct, efficient and expeditious diagnosis of AC and SCC of the lips.

Author Contributions: M.L., A.C. and C.C. contributed to the conception of this study and performed the preliminary documentation. All authors participated in the design of the study and implemented the research. M.L., A.C., C.C., and D.B. were responsible for the data acquisition, selection and analysis, and clinical interpretation of the data. M.L., A.C., and C.C. participated in the statistical analysis and contributed to the interpretation of the results as well as the manuscript drafting and writing of the study. M.L. and C.C. have critically revised the manuscript for important intellectual content. All authors reviewed and approved the final manuscript.

Funding: This research and article processing charges (APC) were funded by a grant of Romanian Ministry of Research and Innovation, CCCDI-UEFISCDI, [project number 61PCCDI/2018 PN-III-P1-1.2-PCCDI-2017-0341], within PNCDI-III.

Acknowledgments: This material is part of the doctoral thesis of Ana Caruntu.

Conflicts of Interest: The authors declare no conflict of interest. The funders had no role in the design of the study; in the collection, analyses, or interpretation of data; in the writing of the manuscript, or in the decision to publish the results.

References

1. De Lucena, E.E.S.; Costa, D.C.B.; da Silveira, E.J.D.; Lima, K.C. Prevalence and factors associated to actinic cheilitis in beach workers. *Oral Dis.* **2012**, *18*, 575–579. [CrossRef]
2. Kaugars, G.E.; Pillion, T.; Svirsky, J.A.; Page, D.G.; Burns, J.C.; Abbey, L.M. Actinic cheilitis: A review of 152 cases. *Oral Surg. Oral Med. Oral Pathol. Oral Radiol Endodontology* **1999**, *88*, 181–186. [CrossRef]
3. Corso, F.; Wild, C.; Gouveia, L.; Ribas, M. Actinic cheilitis: Prevalence in dental clinics from pucpr, curitiba, brazil. *Rev. Clin. Pesq Odontol* **2006**, *2*, 277–281.
4. Markopoulos, A.; Albanidou-Farmaki, E.; Kayavis, I. Actinic cheilitis: Clinical and pathologic characteristics in 65 cases. *Oral Dis.* **2004**, *10*, 212–216. [CrossRef]
5. Cavalcante, A.S.R.; Anbinder, A.L.; Carvalho, Y.R. Actinic cheilitis: Clinical and histological features. *J. Oral Maxillofac. Surg.* **2008**, *66*, 498–503. [CrossRef]
6. Savage, N.W.; McKay, C.; Faulkner, C. Actinic cheilitis in dental practice. *Aust. Dent. J.* **2010**, *55* (Suppl. 1), 78–84. [CrossRef]
7. Vieira, R.A.M.A.R.; Minicucci, E.M.; Marques, M.E.A.; Marques, S.A. Actinic cheilitis and squamous cell carcinoma of the lip: Clinical, histopathological and immunogenetic aspects. *An. Bras. Dermatol.* **2012**, *87*, 105–114. [CrossRef]
8. Nico, M.M.S.; Rivitti, E.A.; Lourenço, S.V. Actinic cheilitis: Histologic study of the entire vermilion and comparison with previous biopsy. *J. Cutan. Pathol.* **2007**, *34*, 309–314. [CrossRef]
9. Miranda, A.M.; Soares, L.G.; Ferrari, T.M.; Silva, D.G.; Falabella, M.E.; Tinoco, E. Prevalence of actinic cheilitis in a population of agricultural sugarcane workers. *Acta Odontol. Latinoam.* **2012**, *25*, 201–207.
10. Miranda, A.M.; Ferrari, T.; Leite, T.; Domingos, T.; Cunha, K.; Dias, E. Value of videoroscopy in the detection of alterations of actinic cheilitis and the selection of biopsy areas. *Med. Oral Patol. Oral Cir. Bucal.* **2015**, *20*, e292–e297. [CrossRef]
11. De Sarmento, D.J.S.; da Miguel, M.C.C.; Queiroz, L.M.; Godoy, G.P.; da Silveira, E.J. Actinic cheilitis: Clinicopathologic profile and association with degree of dysplasia. *Int. J. Dermatol.* **2014**, *53*, 466–472. [CrossRef]
12. Lopes, M.L.; Junior, F.L.S.; Lima, K.C.; Oliveira, P.T.; Silveira, E.J. Clinicopathological profile and management of 161 cases of actinic cheilitis. *An. Bras. Dermatol.* **2015**, *90*, 505–512. [CrossRef]
13. Picascia, D.D.; Robinson, J.K. Actinic cheilitis: A review of the etiology, differential diagnosis, and treatment. *J. Am. Acad. Dermatol.* **1987**, *17*, 255–264. [CrossRef]
14. Ulrich, M.; Gonzalez, S.; Lange-Asschenfeldt, B.; Roewert-Huber, J.; Sterry, W.; Stockfleth, E.; Astner, S. Non-invasive diagnosis and monitoring of actinic cheilitis with reflectance confocal microscopy. *J. Eur. Acad. Dermatol. Venereol.* **2011**, *25*, 276–284. [CrossRef]

15. Cooper, J.S.; Porter, K.; Mallin, K.; Hoffman, H.T.; Weber, R.S.; Ang, K.K.; Gay, E.G.; Langer, C.J. National cancer database report on cancer of the head and neck: 10-year update. *Head Neck* **2009**, *31*, 748–758. [CrossRef]
16. Miranda, A.M.O.; Ferrari, T.M.; Calandro, T.L.L. Queilite actínica: Aspectos clínicos e prevalência encontrados em uma população rural do interior do brasil. *Saúde E Pesquisa* **2011**, *4*, 67–72.
17. Cockerell, C.J. Pathology and pathobiology of the actinic (solar) keratosis. *Br. J. Dermatol.* **2003**, *149*, 34–36. [CrossRef]
18. Holmes, C.; Foley, P.; Freeman, M.; Chong, A.H. Solar keratosis: Epidemiology, pathogenesis, presentation and treatment. *Australas. J. Dermatol.* **2007**, *48*, 67–76. [CrossRef]
19. Wood, N.H.; Khammissa, R.; Meyerov, R.; Lemmer, J.; Feller, L. Actinic cheilitis: A case report and a review of the literature. *Eur. J. Dent.* **2011**, *5*, 101–106. [CrossRef]
20. Kwon, N.H.; Kim, S.Y.; Kim, G.M. A case of metastatic squamous cell carcinoma arising from actinic cheilitis. *Ann. Dermatol* **2011**, *23*, 101–103. [CrossRef]
21. de Abreu, M.A.M.M.; da Silva, O.M.P.; Pimentel, D.R.N.; Hirata, C.H.W.; Weckx, L.L.M.; de Alchorne, M.M.A.; Michalany, N.S. Actinic cheilitis adjacent to squamous carcinoma of the lips as an indicator of prognosis. *Braz. J. Otorhinolaryngol.* **2006**, *72*, 767–771. [CrossRef]
22. Glogau, R.G. The risk of progression to invasive disease. *J. Am. Acad. Dermatol.* **2000**, *42*, S23–S24. [CrossRef]
23. Moy, R.L. Clinical presentation of actinic keratoses and squamous cell carcinoma. *J. Am. Acad. Dermatol.* **2000**, *42*, S8–S10. [CrossRef]
24. Ion, A.; Popa, I.M.; Papagheorghe, L.M.L.; Lisievici, C.; Lupu, M.; Voiculescu, V.; Caruntu, C.; Boda, D. Proteomic approaches to biomarker discovery in cutaneous t-cell lymphoma. *Dis. Markers* **2016**, *2016*, 1–8. [CrossRef]
25. Lupu, M.; Caruntu, C.; Ghita, M.A.; Voiculescu, V.; Voiculescu, S.; Rosca, A.E.; Caruntu, A.; Moraru, L.; Popa, I.M.; Calenic, B.; et al. Gene expression and proteome analysis as sources of biomarkers in basal cell carcinoma. *Dis. Markers* **2016**, *2016*, 1–9. [CrossRef]
26. Voiculescu, V.; Calenic, B.; Ghita, M.; Lupu, M.; Caruntu, A.; Moraru, L.; Voiculescu, S.; Ion, A.; Greabu, M.; Ishkitiev, N.; et al. From normal skin to squamous cell carcinoma: A quest for novel biomarkers. *Dis. Markers* **2016**, *2016*, 1–14. [CrossRef]
27. Solomon, I.; Voiculescu, V.M.; Caruntu, C.; Lupu, M.; Popa, A.; Ilie, M.A.; Albulescu, R.; Caruntu, A.; Tanase, C.; Constantin, C. Neuroendocrine factors and head and neck squamous cell carcinoma: An affair to remember. *Dis. Markers* **2018**, *2018*. [CrossRef]
28. Lupu, M.; Caruntu, A.; Caruntu, C.; Boda, D.; Moraru, L.; Voiculescu, V.; Bastian, A. Non-invasive imaging of actinic cheilitis and squamous cell carcinoma of the lip. *Mol. Clin. Oncol.* **2018**, *8*, 640–646. [CrossRef]
29. Lupu, M.; Căruntu, A.; Moraru, L.; Voiculescu, V.M.; Boda, D.; Tănase, C.; Căruntu, C. Non-invasive imaging techniques for early diagnosis of radiation-induced squamous cell carcinoma of the lip. *Rom. J. Morphol. Embryol.* **2018**, *59*, 949–953.
30. Ridgway, J.M.; Armstrong, W.B.; Guo, S.; Mahmood, U.; Su, J.; Jackson, R.P.; Shibuya, T.; Crumley, R.L.; Gu, M.; Chen, Z.; et al. In vivo optical coherence tomography of the human oral cavity and oropharynx. *Arch. Otolaryngol. Head Neck Surg.* **2006**, *132*, 1074–1081. [CrossRef]
31. Lupu, M.; Popa, I.M.; Voiculescu, V.M.; Boda, D.; Caruntu, C.; Zurac, S.; Giurcaneanu, C. A retrospective study of the diagnostic accuracy of in vivo reflectance confocal microscopy for basal cell carcinoma diagnosis and subtyping. *J. Clin. Med.* **2019**, *8*, 449. [CrossRef] [PubMed]
32. Peppelman, M.; Nguyen, K.P.; Hoogedoorn, L.; van Erp, P.E.J.; Gerritsen, M.J.P. Reflectance confocal microscopy: Non-invasive distinction between actinic keratosis and squamous cell carcinoma. *J. Eur. Acad. Dermatol. Venereol.* **2014**, *29*, 1302–1309. [CrossRef] [PubMed]
33. Guitera, P.; Menzies, S.W.; Longo, C.; Cesinaro, A.M.; Scolyer, R.A.; Pellacani, G. In vivo confocal microscopy for diagnosis of melanoma and basal cell carcinoma using a two-step method: Analysis of 710 consecutive clinically equivocal cases. *J. Invest. Dermatol.* **2012**, *132*, 2386–2394. [CrossRef] [PubMed]
34. Langley, R.G.B.; Walsh, N.; Sutherland, A.E.; Propperova, I.; Delaney, L.; Morris, S.F.; Gallant, C. The diagnostic accuracy of in vivo confocal scanning laser microscopy compared to dermoscopy of benign and malignant melanocytic lesions: A prospective study. *Dermatology* **2007**, *215*, 365–372. [CrossRef] [PubMed]

35. Wolberink, E.A.W.; van Erp, P.E.J.; Teussink, M.M.; van de Kerkhof, P.C.M.; Gerritsen, M.J.P. Cellular features of psoriatic skin: Imaging and quantification using in vivo reflectance confocal microscopy. *Cytom. Part B Clin. Cytom.* **2010**, *80*, 141–149. [CrossRef] [PubMed]
36. Horn, M.; Gerger, A.; Ahlgrimm-Siess, V.; Weger, W.; Koller, S.; Kerl, H.; Samonigg, H.; Smolle, J.; Hofmann-Wellenhof, R. Discrimination of actinic keratoses from normal skin with reflectance mode confocal microscopy. *Dermatol. Surg.* **2008**, *34*, 620–625.
37. Ulrich, M.; Forschner, T.; Röwert-Huber, J.; González, S.; Stockfleth, E.; Sterry, W.; Astner, S. Differentiation between actinic keratoses and disseminated superficial actinic porokeratoses with reflectance confocal microscopy. *Br. J. Dermatol.* **2007**, *156*, 47–52. [CrossRef]
38. Aghassi, D.; Anderson, R.R.; Gonzlez, S. Confocal laser microscopic imaging of actinic keratoses in vivo: A preliminary report. *J. Am. Acad. Dermatol.* **2000**, *43*, 42–48. [CrossRef]
39. Peppelman, M.; Wolberink, E.A.W.; Koopman, R.J.J.; van Erp, P.E.J.; Gerritsen, M.-J.P. In vivo reflectance confocal microscopy: A useful tool to select the location of a punch biopsy in a large, clinically indistinctive lesion. *Case Rep. Dermatol.* **2013**, *5*, 129–132. [CrossRef]
40. Richtig, E.; Ahlgrimm-Siess, V.; Koller, S.; Gerger, A.; Horn, M.; Smolle, J.; Hofmann-Wellenhof, R. Follow-up of actinic keratoses after shave biopsy byin-vivoreflectance confocal microscopy-a pilot study. *J. Eur. Acad. Dermatol. Venereol.* **2010**, *24*, 293–298. [CrossRef]
41. Ulrich, M.; Lange-Asschenfeldt, S.; González, S. In vivo reflectance confocal microscopy for early diagnosis of nonmelanoma skin cancer. *Actas Dermosifiliogr.* **2012**, *103*, 784–789. [CrossRef] [PubMed]
42. Rajadhyaksha, M.; González, S.; Zavislan, J.M.; Rox Anderson, R.; Webb, R.H. In vivo confocal scanning laser microscopy of human skin ii: Advances in instrumentation and comparison with histology. *J. Invest. Dermatol.* **1999**, *113*, 293–303. [CrossRef]
43. Langley, R.G.B.; Burton, E.; Walsh, N.; Propperova, I.; Murray, S.J. In vivo confocal scanning laser microscopy of benign lentigines: Comparison to conventional histology and in vivo characteristics of lentigo maligna. *J. Am. Acad. Dermatol.* **2006**, *55*, 88–97. [CrossRef] [PubMed]
44. González, S.; González, E.; White, W.M.; Rajadhyaksha, M.; Anderson, R.R. Allergic contact dermatitis: Correlation of in vivo confocal imaging to routine histology. *J. Am. Acad. Dermatol.* **1999**, *40*, 708–713. [CrossRef]
45. Ilie, M.A.; Caruntu, C.; Lixandru, D.; Tampa, M.; Georgescu, S.R.; Constantin, M.M.; Constantin, C.; Neagu, M.; Zurac, S.A.; Boda, D. In vivo confocal laser scanning microscopy imaging of skin inflammation: Clinical applications and research directions. *Exp. Ther. Med.* **2019**, *17*, 1004–1011. [CrossRef]
46. Ilie, M.A.; Caruntu, C.; Lupu, M.; Lixandru, D.; Georgescu, S.-R.; Bastian, A.; Constantin, C.; Neagu, M.; Zurac, S.A.; Boda, D. Current and future applications of confocal laser scanning microscopy imaging in skin oncology. *Oncol. Lett.* **2019**. [CrossRef]
47. Ianoși, S.L.; Forsea, A.M.; Lupu, M.; Ilie, M.A.; Zurac, S.; Boda, D.; Ianosi, G.; Neagoe, D.; Tutunaru, C.; Popa, C.M. Role of modern imaging techniques for the in vivo diagnosis of lichen planus. *Exp. Ther. Med.* **2019**, *17*, 1052–1060. [CrossRef]
48. Lupu, M.; Caruntu, C.; Popa, M.I.; Voiculescu, V.M.; Zurac, S.; Boda, D. Vascular patterns in basal cell carcinoma: Dermoscopic, confocal and histopathological perspectives (review). *Oncol. Lett.* **2019**. [CrossRef]
49. Lupu, M.; Popa, I.M.; Voiculescu, V.M.; Caruntu, A.; Caruntu, C. A systematic review and meta-analysis of the accuracy of in vivo reflectance confocal microscopy for the diagnosis of primary basal cell carcinoma. *J. Clin. Med.* **2019**, *8*, 1462. [CrossRef]
50. Caruntu, C.; Boda, D.; Gutu, D.E.; Caruntu, A. In vivo reflectance confocal microscopy of basal cell carcinoma with cystic degeneration. *Rom. J. Morphol. Embryol.* **2014**, *55*, 1437–1441.
51. Ghita, M.A.; Caruntu, C.; Rosca, A.E.; Kaleshi, H.; Caruntu, A.; Moraru, L.; Docea, A.O.; Zurac, S.; Boda, D.; Neagu, M.; et al. Reflectance confocal microscopy and dermoscopy for in vivo, non-invasive skin imaging of superficial basal cell carcinoma. *Oncol. Lett.* **2016**, *11*, 3019–3024. [CrossRef] [PubMed]
52. Rishpon, A.; Kim, N.; Scope, A.; Porges, L.; Oliviero, M.C.; Braun, R.P.; Marghoob, A.A.; Fox, C.A.; Rabinovitz, H.S. Reflectance confocal microscopy criteria for squamous cell carcinomas and actinic keratoses. *Arch. Dermatol.* **2009**, *145*. [CrossRef] [PubMed]
53. Braga, J.C.T.; Scope, A.; Klaz, I.; Mecca, P.; González, S.; Rabinovitz, H.; Marghoob, A.A. The significance of reflectance confocal microscopy in the assessment of solitary pink skin lesions. *J. Am. Acad. Dermatol.* **2009**, *61*, 230–241. [CrossRef] [PubMed]

54. Bağcı, I.S.; Gürel, M.S.; Aksu, A.E.K.; Erdemir, A.T.; Yüksel, E.İ.; Başaran, Y.K. Reflectance confocal microscopic evaluation of nonmelanocytic lip lesions. *Lasers Med. Sci.* **2017**, *32*, 1497–1506. [CrossRef]
55. Georgescu, S.R.; Mitran, C.I.; Mitran, M.I.; Caruntu, C.; Caruntu, A.; Lupu, M.; Matei, C.; Constantin, C.; Neagu, M. Tumour microenvironment in skin carcinogenesis. In *Tumor Microenvironments in Organs*; Springer: Berlin/Heidelberg, Germany, 2020; pp. 123–142.
56. Hartmann, D.; Krammer, S.; Bachmann, M.R.; Mathemeier, L.; Ruzicka, T.; Bagci, I.S.; von Braunmühl, T. Ex vivo confocal microscopy features of cutaneous squamous cell carcinoma. *J. Biophotonics* **2018**, *11*, e201700318. [CrossRef]
57. Ahlgrimm-Siess, V.; Cao, T.; Oliviero, M.; Hofmann-Wellenhof, R.; Rabinovitz, H.S.; Scope, A. The vasculature of nonmelanocytic skin tumors on reflectance confocal microscopy. *Arch. Dermatol.* **2011**, *147*, 264. [CrossRef]
58. Skobe, M.; Rockwell, P.; Goldstein, N.; Vosseler, S.; Fusenig, N.E. Halting angiogenesis suppresses carcinoma cell invasion. *Nat. Med.* **1997**, *3*, 1222–1227. [CrossRef]

© 2020 by the authors. Licensee MDPI, Basel, Switzerland. This article is an open access article distributed under the terms and conditions of the Creative Commons Attribution (CC BY) license (http://creativecommons.org/licenses/by/4.0/).

MDPI
St. Alban-Anlage 66
4052 Basel
Switzerland
Tel. +41 61 683 77 34
Fax +41 61 302 89 18
www.mdpi.com

Journal of Clinical Medicine Editorial Office
E-mail: jcm@mdpi.com
www.mdpi.com/journal/jcm

www.ingramcontent.com/pod-product-compliance
Lightning Source LLC
LaVergne TN
LVHW070551100526
838202LV00012B/441